TREES, SHRUBS, & CACTI

of South Texas

TREES, SHRUBS, & CACTI
of South Texas

James H. Everitt and D. Lynn Drawe

Texas Tech University Press

This book was set in 10 on 13 Palatino and printed on acid-free paper that meets the guidelines for permanence and durability of the Committee on Production Guidelines for Book Longevity of the Council on Library Resources.

Cover design by Kerri Carter
Book design by Kelley Ferguson Farwell

Manufactured in Singapore by Palace Press

Library of Congress Cataloging-in-Publication Data

Everitt, J. H. 1944–
Trees, shrubs & cacti of south Texas / James H.
Everitt and D. Lynn Drawe.
p. cm.
Includes bibliographical references.
ISBN 0-89672-252-X. — ISBN 0-89672-253-8 (pbk.)
1. Trees—Texas—Identification. 2. Shrubs—Texas—Identification.
3. Cactus—Texas—Identification. 4. Trees—Texas—Pictorial works.
I. Drawe, Dale Lynn, 1942– . II. Title. III. Title: Trees, shrubs, and
cacti of south Texas.
QK188.E83 1992
582.1609764—dc20

91-32009
CIP

Texas Tech University Press
Lubbock, Texas 79409-1037 USA

94 95 96 97 98 99 00 01 / 9 8 7 6 5 4 3 2

CONTENTS

This book is dedicated to the memory of Doris Garrison Drawe, 1921–1972

PREFACE

This publication contains a color photograph, the family name, scientific name, common name, general description, geographical range, and ecological characteristics for each species of the majority of trees, shrubs, and cacti occurring in the 14 southernmost counties of Texas. It does not include every species of tree, shrub, or cactus known to occur there, but it does include frequently encountered species as well as many of the rare or endangered species.

Although this publication covers only the 14 southernmost counties in Texas, the extensive ranges of many of the represented species make it a useful reference for plants in other areas of Texas, the southwestern United States, and northern Mexico. In general, most plants were photographed at either the flowering or fruiting stage so that they might be more easily recognized.

Taxonomists classify plants as herbs, subshrubs, shrubs, and trees. An *herb* is a plant without a persistent woody stem, at least above ground; a *subshrub* is a perennial plant with the lower portion of the stems woody and persistent; a *shrub* is a woody plant smaller in size than a tree, which usually produces several branches from the base; a *tree* is a woody plant with one main stem and is at least four or five meters tall. In this publication we have included many subshrub species along with the numerous species of shrubs and trees.

Some 190 species and 57 families are represented, including 22 species of cacti and 168 species of trees, shrubs, and subshrubs. Nearly all of the plants represented are native, but some naturalized, introduced plants are also represented. A native plant is indigenous to an area and grows without cultivation, whereas a naturalized plant has escaped from cultivation and reproduces itself in the wild. Most of the trees, shrubs, and subshrubs included here are represented by herbarium specimens kept at the USDA-ARS Subtropical Texas Area Location in Weslaco, Texas and at the Welder Wildlife Foundation Refuge at Sinton, Texas. The cacti are not represented by herbarium specimens.

This publication will be useful to ranchers and ranch managers, scientists, and anyone who is interested in the flora of southern Texas. It will enable them to identify many of the plants in this area. Information provided will be useful to help develop sound land management programs.

Further information about the woody plants and cacti of this area can be found in Donovan S. Correll and Marshall C. Johnston's 1970 *Manual of the Vascular Plants of Texas*; Robert A Vines' 1960 *Trees, Shrubs, and Woody Vines of the Southwest*; and Del Weniger's 1970 *Cacti of the Southwest*.

Plant Names, Descriptions, and Geographical Ranges

Scientific and most common names for trees, shrubs, and subshrubs used herein are according to Correll and Johnston and Vines. Some of the Spanish common names were obtained from Elias J. Guerrero's checklist on *Scientific, Standard, and Spanish Names of Woody Plants in South Texas,* and from Robert Runyon's *Vernacular Names of Plants Indigenous to the Lower Rio Grande Valley of Texas.* Scientific and common names used for cacti follow the nomenclature of Weniger. Information on plant descriptions and geographical ranges of trees, shrubs, and subshrubs was obtained from Correll and Johnston and Vines, whereas information on descriptions and geographical ranges of cacti was taken from Weniger. Notes concerning wildlife, livestock, and human values were taken from the authors' research and personal knowledge.

Acknowledgments

The authors thank Norman A. Browne, University of Texas–Pan American, Edinburg, Texas, for preparing the leaf illustrations and Jeanne Everitt for her encouragement and assistance in obtaining the photographs. Thanks are also extended to Mario Alaniz and Wayne Swanson for preparation of figures and help in obtaining photographs, and to Saida Cardoza and Ofelia de la Fuente for typing the manuscript. We wish to express our gratitude to Robert I. Lonard, University of Texas–Pan American, and George Williges, Texas A&I University, for reviewing drafts.

This publication was partially funded by a grant from the Rob and Bessie Welder Wildlife Foundation and is Welder Wildlife Foundation Contribution Number B-10.

INTRODUCTION

The southernmost counties of Texas (see map) are a part of the South Texas or Rio Grande Plains and the Gulf Prairies and Marshes vegetational areas (Gould, 1975). The area is part of the Tamaulipan Biotic Province. These counties account for nearly 11 million acres (Texas Almanac, 1986) bordered by Mexico on the west and the Gulf of Mexico on the east.

The topography of this area is level to rolling, and the land is dissected by numerous arroyos and ephemeral streams that flow into the Rio Grande or the Gulf of Mexico. Elevations range from sea level to 800 feet (Texas Almanac, 1986).

The climate of the area is mild with short winters and relatively warm temperatures throughout the year. The growing season varies from 300 days in San Patricio County in the northeast portion of the area to nearly 365 days in Cameron County in the extreme southern portion known as the Lower Rio Grande Valley. Average annual rainfall ranges from 43 to 76 centimeters (Texas Almanac, 1986), increasing from west to east. The lowest long-term average rainfall occurs in January and February, whereas the highest occurs in May or June and again in September after a midsummer depression. Rainfall may come as local showers or as high-intensity rains. Rainfall of over 25 centimeters in a single day has been recorded in the area. Periodic droughts often occur and there are months when no precipitation occurs. Evaporation rates exceed precipitation severalfold, especially in western portions of this area.

Soils range from acid sands, clay loams, and saline clays along coastal areas to clay, sand, and sandy loams inland. They vary from very basic to slightly acidic. A wide range of soil profile types causes great differences in soil drainage and moisture holding capacities. Typical range sites include deep sands, hardlands, shallow ridges, bottomlands, alkali flats, mixed sandy land, coastal flats, and marshes (Correll and Johnston, 1970).

Although there are large acreages of cultivated land in this area, the majority of the land holdings are large cattle ranches. The area is rich in wildlife, particularly white-tailed deer. The economic value of these animals is an important consideration in vegetation management.

The flora of this area is complex because of the wide variability in climate, soils, and topography. Although the flora consists primarily of herbaceous species (grasses and flowering herbs), the area contains many trees, shrubs, subshrubs, and cacti. Because a sound land management program cannot be developed without knowledge of the plants growing in an area, the rancher, land manager, or wildlife biologist needs to know the flora and its attributes.

The several kinds of plants in the southernmost Texas area have many different uses and characteristics. Some are poisonous, some provide food for livestock and wildlife, some furnish a home or cover for wildlife, some provide nectar for honey-producing bees, some provide erosion control, and some are used as ornamentals. The main contribution of some plants is simply the spectacular, unspoiled beauty they provide. Unfortunately, some plants in this area are endangered and an effort needs to be made for their conservation to maintain the unique flora of southern Texas.

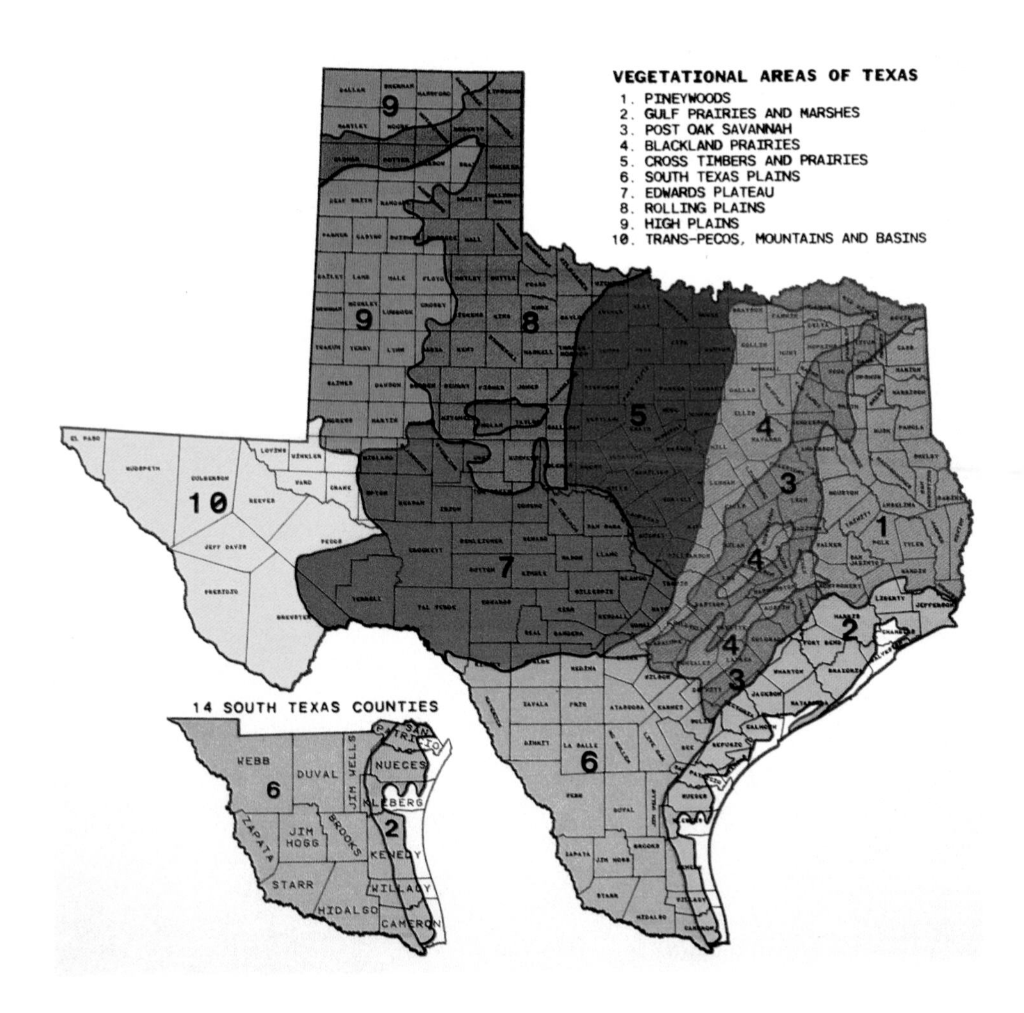

VEGETATIONAL AREAS OF TEXAS
1. PINEYWOODS
2. GULF PRAIRIES AND MARSHES
3. POST OAK SAVANNAH
4. BLACKLAND PRAIRIES
5. CROSS TIMBERS AND PRAIRIES
6. SOUTH TEXAS PLAINS
7. EDWARDS PLATEAU
8. ROLLING PLAINS
9. HIGH PLAINS
10. TRANS-PECOS, MOUNTAINS AND BASINS
14 SOUTH TEXAS COUNTIES
WEBB
DUVAL
JIM WELLS
SAN PATRICIO
NUECES
KLEBERG
ZAPATA
JIM HOGG
BROOKS
KENEDY
STARR
WILLACY
HIDALGO
CAMERON

ACANTHACEAE

CARLOWRIGHTIA

Carlowrightia parviflora (Buckl.) Wasshausen

Small subshrub attaining a height of 45 cm with slender, branching, ascending stems. Leaves opposite, oblong to lanceolate, with entire margins that are rolled under. Flowering February–June; flowers light blue or white; fruit a four-seeded capsule.

Infrequent to rare in gravelly or sandy loam soils in chaparral of the Rio Grande Plains (Starr County).

TETRAMERIUM

Tetramerium platystegium Torr.

Low, slender, much-branched subshrub usually 2–4 dm tall. Leaves simple, opposite, elliptic-lanceolate to elliptic-linear, with entire margins. Flowering in summer; flowers lavender; fruit a small, two-seeded capsule. The prominent yellow bracts are a conspicuous characteristic of this plant.

Rare on rocky hills in the Rio Grande Plains (Starr County) and Edwards Plateau.

AMARYLLIDACEAE

CENTURY PLANT, MAGUEY
Agave americana L.

Medium to large, acaulescent plant with a basal rosette of large, gray leaves. Plants attain a height of 1.5–2 m and may be 2 m broad. Leaves 12–17 dm long with straight marginal spines. Flowering June–July; flowers yellow; fruit a many-seeded, brown capsule.

Found throughout the southern Rio Grande Plains where it is widely cultivated, has escaped to the wild, and has become naturalized.

Plants live many years but bloom only once, then die.

THORN-CRESTED AGAVE
Agave lophantha Schiede

Small, rather open, acaulescent plant with a basal rosette of light green leaves. This plant attains a maximum height of 3–4 dm. The leaves are 3–4 dm long and have variously curved, marginal spines. Flowering in late spring; flowers pale green to yellowish green; fruit a many-seeded, brown capsule.

Infrequent to rare on sandy loam soils in the extreme southwestern Rio Grande Plains (Starr and Zapata counties).

ROUGH AGAVE, MAGUEY CENIZO

Agave scabra Salm-Dyck

[syn. *A. asperrima* Jacobi]

Medium to large, acaulescent plant with an open basal rosette of large, gray leaves. The leaves are characterized by having a very rough surface. Plants attain a maximum height of about 1.5 m and may be 1.5 m broad. The leaves are 7–11 dm long and heavily armed with strong, reflexed, marginal spines. Flowering April–June; flowers yellow; fruit a many-seeded, brown capsule.

On sandy and calcareous soils in the extreme southwestern Rio Grande Plains (Starr, Webb, and Zapata counties).

This species lives many years but blooms only once, then dies.

APOCYNACEAE

ROCK-TRUMPET, FLOR DE SAN JUAN

Macrosiphonia macrosiphon (Torr.) Heller

Erect or somewhat diffuse herb with a woody base, attaining a height of 1.5–3 dm. Leaves opposite, ovate-elliptic with slightly toothed margins. The leaves are densely pubescent on both sides. Flowering May–September; flower a showy white or pink-tinged trumpet; fruit a many-seeded follicle.

Found on dry, rocky, open or chaparral-covered slopes in the extreme southern Rio Grande Plains (Hidalgo and Starr counties), western Edwards Plateau, and Trans-Pecos.

AQUIFOLIACEAE

YAUPON
Ilex vomitoria Ait.

Evergreen shrub or tree to 8 m high with short, stout, rigid twigs. Leaves simple, alternate, elliptic to oblong or oblong-elliptic to ovate-elliptic, with toothed margins. Flowering April–May; flowers white; fruit a red drupe.

In the Pineywoods, Coastal Prairie (southwest along the coast to San Patricio and Nueces counties), and Edwards Plateau.

The fruit is eaten by several species of passerine birds including robins and cedar waxwings. It is also consumed by raccoons and opossums.

AVICENNIACEAE

BLACK-MANGROVE, MANGLE BLANCO
Avicennia germinans (L.) L.

Shrub or small tree rarely over 1 m tall. Leaves simple, opposite, oblong or lanceolate to elliptic or obovate, with entire margins. The leaves are shiny above and pale pubescent beneath. Flowering July–August; flowers creamy white; fruit usually a one-seeded, two-valved capsule.

On sandy or clay tidal flats and lagoons along the Coastal Prairies and Marshes of south and southeast Texas.

BATACEAE

VIDRILLOS, MARITIME SALTWORT
Batis maritima L.

Low, pale green shrub with creeping stems. Leaves thick, succulent, opposite, linear or boat-shaped, with entire margins. Flowering June–August; flowers very small and whitish, male and female flowers separate on the same plant; fruit a fleshy aggregate.

On sandy beaches, mud flats, and saline marshes along the Coastal Prairies and Marshes of south Texas.

BERBERIDACEAE

AGARITO, WILD CURRANT, CHAPARRAL BERRY
Berberis trifoliolata Moric.

Evergreen shrub attaining a height of 2 m with stiff, spiny, hollylike, trifoliolate, alternate leaves. Flowering February–April; flowers yellow; fruit a red berry with one to several seeds.

Found on rocky slopes, thickets, and open woods from coastal south Texas (Kleberg, Nueces, and San Patricio counties) northwest and westward into the Trans-Pecos.

The red berries make an excellent jelly. Many birds and mammals are known to eat the fruit including cardinals, mockingbirds, raccoons, opossums, gray foxes, and coyotes. The leaves are browsed by white-tailed deer. The reddish new growth is quite tart and tasty and makes a unique garnish for salads.

BORAGINACEAE

ANACAHUITA, MEXICAN OLIVE

Cordia boissieri A. DC.

Shrub or small tree to 8 m tall with a spreading, rounded top and stout branches. Leaves simple, alternate, covered with dense pubescence, ovate to lance-ovate, with entire margins. Flowering throughout the year; flowers white, very showy; fruit a white to yellow green drupe (turning brown with age).

On dry soil in the southern Rio Grande Plains.

This species is widely planted as an ornamental. A jelly made from the fruit is used as a remedy for coughs and colds in Mexico. The fruit is edible and sweet and is eaten by cattle, white-tailed deer, and javelina. Several species of birds use this species as nesting sites.

ANAQUA, SUGARBERRY, KNOCK-AWAY

Ehretia anacua (Teran & Berl.) I. M. Johnst.

Shrub or tree with a rounded canopy, attaining a height of 15 m. Leaves simple, alternate, elliptic to ovate or less commonly, broadly lanceolate, with entire or partially toothed margins. The leaf surface is rough to the touch. Flowering June–October; flowers white; fruit a yellow to orange, two-seeded drupe.

Found in thickets, palm groves, and along fencerows and chaparral from the southern Edwards Plateau south through the Rio Grande Plains.

A number of mammals and birds eat the fruit including the coyote, raccoon, and chachalaca. It is often planted as an ornamental. White-tailed deer occasionally browse the leaves.

CROWDED HELIOTROPE

Heliotropium confertifolium (Torr.) Gray

Low, perennial plant usually less than 1 dm tall with a woody base and taproot. Stems and leaves covered with silky, silvery pubescence. Leaves alternate, linear, small, and crowded. Flowering March–October; flowers white; fruit four-lobed, separating into four, one-celled nutlets.

On dry soils of the Trans-Pecos and western Rio Grande Plains (Hidalgo, Jim Hogg, Jim Wells, Starr, Webb, and Zapata counties).

NARROW-LEAF HELIOTROPE

Heliotropium torreyi I. M. Johnst.

Small, gray green shrub with numerous erect branches, attaining a height of 5 dm. This plant is characterized by having all parts densely covered with appressed hairs. Leaves numerous, alternate, linear to almost filiform, covered with white hairs. Flowering April–November; flowers yellowish white; fruit a small, four-lobed capsule containing four nutlets.

On caliche or gravelly hillsides in the Rio Grande Plains (Duval, Jim Hogg, Jim Wells, and Hidalgo counties), Trans-Pecos, and Edwards Plateau.

This small shrub is reported to be a good browse for sheep.

OREJA DE PERRO
Tiquilia canescens (A. DC.) A. Richards
[syn. *Coldenia canescens* DC.]

Low, much-branched plant from a woody taproot, rarely more than 3 dm tall with all parts densely pubescent. Leaves simple, alternate, ovate to elliptic or elliptic-lanceolate, with entire margins that are partially rolled inward at the edge. Flowering March–August; flowers pink, rose, or rarely white; fruit a small nutlet.

On limestone or calcareous soils from the Trans-Pecos south through the western Rio Grande Plains.

BROMELIACEAE

GUAPILLA
Hechtia glomerata Zucc.

Plant to 18 dm tall with the habit of *Agave* or *Yucca*, with practically no stem or if present, a very short one. Leaves arranged in a densely spreading rosette, linear-lanceolate, usually recurving with stout, recurved spines. Flowering May–August; male and female flowers on separate plants; fruit a many-seeded, brown capsule.

On gravelly sites and sandstone formations in Starr and Zapata counties in the extreme southern Rio Grande Plains.

CACTACEAE

NIGHT BLOOMING CEREUS, ORGANO, BARBED-WIRE CACTUS
Acanthocereus pentagonus (L.) Britt. & Rose

Semierect or erect cactus that leans or grows over other plants, usually 1–2 m tall. This cactus forms thickets by the arching 3-to-5 angled stems that root where they touch the ground. Flowering at night June–August; flowers white; fruit an oblong, fleshy, many-seeded, red berry.

Found in coastal portions and the extreme southern Rio Grande Plains (Cameron, Hidalgo, Kenedy, and Willacy counties).

The fruit is eaten by several species of birds and mammals including the bobwhite quail, white-tailed deer, javelina, and raccoon.

STAR CACTUS, SEA-URCHIN CACTUS, STAR PEYOTE
Echinocactus asterias Zucc.

Extremely flat, depressed, disk-shaped to sometimes low, dome-shaped cactus reaching a maximum diameter of 15 cm. Flowering spring–fall; flowers yellow; fruit an oval berry covered with dense hairs.

A rare cactus, found at a few locations in Starr County in the extreme southern portion of the Rio Grande Plains.

GLORY OF TEXAS

Echinocactus bicolor var. *schottii* Eng.

Egg-shaped or conical to almost columnar-shaped cactus attaining a maximum height of 25 cm and a width of 12 cm. Flowering spring–summer; flowers pink, red, or purple; fruit a berry.

This cactus occurs in two widely separated areas, Starr County in the Rio Grande Plains and the Big Bend area of the Trans-Pecos.

FISHHOOK CACTUS, ROOT CACTUS

Echinocactus scheeri SD.

Columnar- or club-shaped cactus attaining a height of 20 cm. This cactus is unique in having a long, fleshy, white taproot. The spines are usually shaped like a fishhook. Flowering in early spring; flowers pale green or yellow green; fruit a green, club-shaped berry.

Usually on gravelly hillsides throughout the Rio Grande Plains.

FISHHOOK CACTUS, HEDGEHOG CACTUS, TWISTED-RIB CACTUS
Echinocactus setispinus Eng.

Columnar-shaped cactus reaching a maximum height of 30 cm but usually smaller. This cactus sometimes has spines shaped like a fishhook. Flowering spring–summer; flowers yellow; fruit a red berry.

Found throughout the Rio Grande Plains, Edwards Plateau, and the eastern Trans-Pecos.

LOWER RIO GRANDE VALLEY BARREL CACTUS
Echinocactus sinuatus Dietrich

Spherical, becoming conical and finally elongated-shaped cactus growing to 30 cm tall and 20 cm wide. Flowering in summer; flowers lemon yellow; fruit a green, egg-shaped berry.

Found from the lower Rio Grande Valley in a narrow strip north along the Rio Grande in the western Rio Grande Plains.

HORSE CRIPPLER, DEVIL'S HEAD, DEVIL'S PINCUSHION, MANCA CABALLO

Echinocactus texensis Höpffer

Spherical-shaped cactus to 3 dm in diameter. Flowering in early spring; flowers pink to red orange; fruit a many-seeded, red berry.

A frequent species found in the Trans-Pecos, Edwards Plateau, Rio Grande Plains, and southern Coastal Prairie.

This cactus is regarded as a menace to cattlemen because unsuspecting stock animals can be badly injured by the strong thorns. The fruits are edible and tasty.

BERLANDIER'S ALICOCHE

Echinocereus berlandieri (Eng.) Rumpl.

Sprawling, clustering, and branching cactus with the older stems lying on the ground but with the growing tips and sometimes complete stems erect. The stems attain a maximum height of 15 cm. Flowering March–April; flowers red-pink with whitish centers; fruit a green, egg-shaped berry.

Found throughout most of the Rio Grande Plains but most common along the Nueces River and lower Rio Grande.

ALICOCHE

Echinocereus blanckii (Poselgr.) Palmer

Low cactus with soft, wrinkled, twisted stems that attain a maximum length of 30 cm. Flowering in spring; flowers rose-colored in the upper parts with dark red centers; fruits unknown.

Found along the Rio Grande in the extreme southern Rio Grande Plains (Hidalgo and Starr counties).

STRAWBERRY CACTUS, PITAYA

Echinocereus enneacanthus Eng.

Clump-forming cactus with many stems that may be up to 3 dm tall. These stems grow in loose clusters of a few to as many as 100 in a large plant. Flowering in spring (usually April); flowers very large and showy, purple red in color; fruit a round berry, greenish to brownish or purplish.

Found in the western Rio Grande Plains and westward to the Big Bend area.

The flesh of the fruit is edible, delicious, and has a flavor similar to strawberries.

ECHINOCEREUS

Echinocereus fitchii B. & R.

Upright, usually single-stemmed cactus attaining a height of 15 cm. Flowering in spring; flowers large and showy, pink with dark burgundy centers; fruit an oval, green berry covered with white, dense hair.

On rocky hills overlooking the Rio Grande in the western Rio Grande Plains (Starr, Webb, and Zapata counties).

SMALL PAPILLOSUS

Echinocereus papillosus var. *angusticeps* (Glover) Marshall

Small columnar-shaped cactus attaining a height of 10 cm. This cactus often occurs in dense clusters with many stems. Flowering in early spring; flowers large, showy, and greenish yellow; fruit a greenish berry covered with short bristles.

Found only in northern Hidalgo County in the extreme southern Rio Grande Plains.

PEYOTE

Lophophora williamsii (Lem.) Coult.

Small, spherical-shaped, spineless cactus from 5–10 cm in diameter. Flowering spring–fall; flowers pale pink; fruit a small, red berry with black seeds.

On calcareous or limestone soils from Presidio County in west Texas south along the Rio Grande to Starr and Jim Hogg counties in the Rio Grande Plains.

The plant is narcotic and has been used by Indians in their religious ceremonies to produce hallucinations.

NIPPLE CACTUS, BIZNAGA DE CHILITOS, LITTLE CHILIS

Mammillaria heyderi Muehlenpf.

Spherical-shaped cactus up to 12 cm in diameter and 5 cm in height. Flowering in spring; flowers brownish, pinkish, or very pale purple shading to white; fruit a red, club-shaped berry.

Found throughout the Rio Grande Plains and Trans-Pecos.

HAIR-COVERED CACTUS, GRAPE CACTUS

Mammillaria multiceps SD.

Small, spherical- to egg-shaped cactus attaining a height of approximately 6 cm. It is easily recognized by the flexible, hairlike, radial spines that cover its surface. Flowering in spring; flowers brownish yellow or almost tan; fruit a many-seeded, egg-shaped to club-shaped, red berry.

Found in the extreme southern (lower Rio Grande Valley) and western Rio Grande Plains and the southwestern Edwards Plateau.

RUNYON'S ESCOBARIA, JUNIOR TOM THUMB CACTUS

Mammillaria roberti Berger

Small, oblong-shaped cactus attaining a height of 8 cm with stiff, white, brown-tipped spines. This cactus often forms low clumps of dozens of stems. Flowering in spring; flowers inconspicuous, buff or tan; fruit a spherical- to egg-shaped, red berry.

On gravelly hillsides in the western Rio Grande Plains.

RUNYON'S CORYPHANTHA, DUMPLING CACTUS
Mammillaria runyonii B. & R.

Irregularly sized and shaped cactus having a large, succulent, carrot-shaped taproot. The stems form masses to nearly 50 cm in diameter. Flowering in spring; flowers purplish or rose pink; fruit a green berry.

Found in the extreme southern Rio Grande Plains (Cameron and Starr counties).

MAMMILLARIA
Mammillaria sphaerica Dietrich

Low, spherical-shaped cactus usually consisting of a low mass up to 30 cm in diameter. This cactus has a thick, fleshy taproot. Flowering spring–summer; flowers lemon yellow; fruit an egg-shaped berry.

Found in the southern Rio Grande Plains, usually growing in shade.

TEXAS PRICKLY PEAR, CACANAPO, NOPAL

Opuntia engelmannii SD.

[syn. *O. lindheimeri* Engelm.]

Thicket-forming, heavy-bodied cactus attaining a height of 1–3 m. Flowering April–June; flowers yellow, orange, or red; fruit a many-seeded, red or purple berry.

Found in central and southeast Texas and south through the Rio Grande Plains and southern Coastal Prairie.

The fruit may be eaten raw or made into a preserve, and the joints (nopalitos) when young and tender are cooked and served with a peppery dressing. During droughts, ranchers burn off the spines to provide emergency forage for cattle. The joints and fruit are eaten by white-tailed deer, javelina, and many other species of wildlife. The plants are also used as nesting sites by the cactus wren. This species is highly important to animals not only as a source of food and cover but also as a source of water during drought.

DESERT CHRISTMAS CACTUS, TASAJILLO, TURKEY PEAR

Opuntia leptocaulis DC.

Thicket-forming cactus usually bushy, erect or sometimes reclining on other plants, 3–18 dm tall. Flowering mostly July–August; flowers green to yellow or bronze and inconspicuous; fruit a bright red, juicy, many-seeded berry.

Most common on heavier or bottomland soils in the Trans-Pecos, Edwards Plateau, southern Coastal Prairie, and Rio Grande Plains.

The red fruits are eaten by Rio Grande turkey, and both the fruits and joints are consumed by white-tailed deer.

DEVIL CACTUS, DOG CHOLLA, CLAVELLINA
Opuntia schottii Eng.

Mat-forming cactus, the joints elongated, cylindrical- or club-shaped. Flowering in spring; flowers yellow; fruit a many-seeded, yellow berry.

It grows on gravelly hillsides overlooking the Rio Grande in the western Rio Grande Plains and Trans-Pecos.

RAT-TAIL CACTUS, WILCOX CACTUS
Wilcoxia poselgeri (Lemaire) Britt. & Rose

An infrequent, inconspicuous cactus attaining a height of 3–9 dm with slender, erect, or reclining stems. This species usually grows entangled with other thorny shrubs in thickets and is easily overlooked. Flowering February–April; flowers pink, very showy; fruit a many-seeded, pubescent berry.

On dry, gravelly or sandy hillsides from Hidalgo to Webb County in the western Rio Grande Plains.

CAPPARIDACEAE

VARA BLANCA
Capparis incana H.B.K.

Rare evergreen shrub or tree covered with grayish or brownish hairs. Leaves simple, alternate, linear with entire margins. Flowering spring–fall; flowers white; fruit globose or oblong.

Known from two locations in the extreme southern Rio Grande Plains (Cameron and Hidalgo counties).

CELASTRACEAE

GUTTA-PERCHA, LEATHERLEAF, MANGLE DULCE
Maytenus texana Lundell

Much-branched, creeping evergreen shrub or small tree attaining a height of 1–2 m. Leaves simple, fleshy, alternate, oblong-elliptic to elliptic-obovate, with entire or remotely toothed margins. Flowering January–July; flowers very small, yellowish green; fruit a small, two- to four-seeded, red capsule.

Found in thickets on clay and sandy clay mounds and rises in the Coastal Prairies and Marshes of south Texas (Cameron, Kenedy, Kleberg, Nueces, and Willacy counties).

AFINADOR, GREGG MORTONIA
Mortonia greggii Gray

Low, branching, erect shrub with pubescent stems, attaining a height of 2 m. Leaves evergreen, crowded, alternate, oblanceolate with entire margins. Flowering December–May; flowers white; fruit a small, one-seeded capsule.

Infrequent in thickets on limestone and caliche hills in the western Rio Grande Plains (Hidalago and Starr counties).

DESERT YAUPON, CAPUL, PANALERO
Schaefferia cuneifolia Gray

Densely branched, rigid, evergreen shrub usually 1–2 m tall with somewhat spinescent twigs. Leaves simple, alternate, cuneate-obovate to oblanceolate, with entire margins. Flowering February–September; flowers light green to white, male and female flowers on separate plants; fruit a one- or two-seeded, orange to red drupe.

Found on rocky hillsides and in chaparral in the Trans-Pecos and Rio Grande Plains; occasional on more xeric sites in the southern Coastal Prairie.

The leaves of this shrub are browsed by white-tailed deer, whereas the fruits are eaten by several species of birds and mammals including the bobwhite quail, scaled quail, cactus wren, coyote, and woodrat.

ARMED SALTBUSH, TUBERCLED SALTBUSH, HUAHA
Atriplex acanothocarpa (Torr.) Wats.

Evergreen subshrub attaining a height of 3–10 dm, herbaceous above, woody and branched from the base. Leaves alternate or the lowest opposite, silver-gray, oblong to oblong-lanceolate, with entire or toothed margins. Flowering summer–fall; male and female flowers on separate plants; bracts of the fruit conspicuous with numerous, flattened tubercles.

Found on alkaline soils in the western half of Texas, infrequent in the extreme southern Rio Grande Plains (Cameron, Starr, Webb, and Zapata counties).

This species is sometimes used for windbreaks, roadside cover, and as an ornamental. It is occasionally browsed by cattle.

FOUR-WING SALTBUSH, WING-SCALE, SHADSCALE
Atriplex canescens (Pursh) Nutt.

Erect, evergreen, loosely to densely branched shrub 1–2 m tall. Leaves numerous, alternate, silver-gray, linear-spatulate to narrowly-oblong, with entire margins. Flowering April–October; male and female flowers on separate plants, rarely on the same plant; fruit a four-winged bract.

Found on dry mesas and alkaline valleys, prairies, and hillsides throughout the western half of Texas; an infrequent species in the western half of the Rio Grande Plains.

This plant is a palatable and nutritious feed for cattle, sheep, and goats. It is very drought resistant and is useful for erosion control.

MEXICAN DEVIL-WEED, DEVIL-WEED ASTER

Aster spinosus Benth.

Shrublike plant attaining a height of 18 dm, herbaceous above with a woody base. This plant is characterized by having very erect, slender, green branchlets. The axillary branches are often transversed into spines. Leaves often absent, when present they are inconspicuous, alternate, and linear-spatulate in shape. Flowering summer–fall; flowers yellowish white; fruit an achene.

Locally abundant on roadsides and other weedy slopes and banks in the Coastal Prairies and Marshes, Rio Grande Plains, and Trans-Pecos.

White-tailed deer browse the new leaves of this plant in spring and its flowers in the fall.

ROOSEVELT WILLOW, NEW DEAL WEED, JARA DULCE, DRY-LAND WILLOW, FALSE WILLOW

Baccharis neglecta Britt.

Much-branched shrub 1–3 m tall with ascending branches. Leaves simple, alternate, narrowly linear to narrowly elliptic, with entire or remotely toothed margins. Flowering summer–fall; flowers white, male and female flowers on separate plants; fruit an achene.

Found nearly throughout the state except for the dense forest of the east Texas Pineywoods and the higher elevations of the High and Rolling Plains. This shrub is often found in disturbed habitats. It is similar to seepwillow but its leaves are much narrower and the flowers are smaller.

JARA, SEEPWILLOW, WATER-WALLY

Baccharis salicifolia (R. & P.) Pers.

Shrub of stream banks forming thickets in clumps, usually 1–3 m tall. Leaves simple, alternate, lanceolate to narrowly elliptic, with toothed margins. Flowering summer–fall; flowers white; fruit an achene. The male and female flowers are on separate plants.

Found along sandy watercourses in dry areas of the Trans-Pecos, Edwards Plateau, and Rio Grande Plains (Cameron, Hidalgo, Starr, and Zapata counties).

SEA OX-EYE

Borrichia frutescens (L.) DC.

Small, whitened shrub usually 4–8 dm (rarely 12 dm) tall with thick and somewhat fleshy, gray green leaves. Leaves simple, opposite, obovate to oblanceolate to spatulate, with entire or partially toothed margins. Flowering throughout the year; flowers yellow; fruit an achene.

Very abundant in Coastal Prairies and Marshes of south Texas and inland in local areas of poor drainage and salt accumulation in the Rio Grande Plains (Hidalgo, Starr, and Webb counties).

CLAPPIA

Clappia suaedaefolia Gray

Small subshrub attaining a height of 6 dm, mostly fleshy and herbaceous but much-branched and woody near the base. Leaves opposite on the lower part of the stem but mostly alternate, fleshy, and linear with entire margins. The leaves are confined to the lower two-thirds or three-fourths of the plant. Flowering spring–fall; flowers yellow; fruit an achene.

Infrequent to rare on subsaline, poorly drained clay flats of the Rio Grande Plains (Cameron, Hidalgo, and Starr counties).

FALSE BROOMWEED

Ericameria austrotexana M. C. Johnst.

Erect, rounded, much-branched shrub 5–15 dm tall. Leaves numerous, alternate, linear with entire margins. Flowering summer–fall; flowers pale yellow; fruit an achene.

Frequent in open ground along the Coastal Prairies and Marshes from southeast Texas southwest and south to Cameron County. Also found inland in the Rio Grande Plains (Hidalgo, Jim Hogg, Starr, and Zapata counties).

BLUE EUPATORIUM, BLUE BONESET

Eupatorium azureum DC.

Much-branched, scandent shrub or woody, nontwining vine attaining a height of 2–3 m. Leaves opposite (those of branches smaller and narrower than those of stem), deltoid with coarsely toothed margins. Flowering February–May; flowers bluish or blue lavender; fruit an achene.

Local in dense thickets in the extreme southern Rio Grande Plains (Cameron, Hidalgo, and Willacy counties).

CRUCITA

Eupatorium odoratum L.

Subshrubby plant attaining a height of 1–2 m with erect or reclining branches that are herbaceous above and woody near the base. Leaves opposite, deltoid or rhombic-ovate, with entire or sparsely toothed margins. The leaves of the branches are smaller and narrower than those of the stem. Flowering late summer–fall; flowers lilac to bright purplish blue; fruit an achene.

Frequent in the Coastal Prairie of southeast Texas and the Rio Grande Plains (Cameron, Hidalgo, Jim Wells, Kleberg, and Nueces counties). Usually found on rocky or clay soils.

Frequently consumed by white-tailed deer.

OCOTE, CHOMONQUE
Gochnatia hypoleuca DC.

Much-branched shrub usually 1–2 m tall with twigs and bottom sides of leaves pale-colored and covered with densely matted pubescence. The upper surfaces of the leaves are dark, almost blackish green. Leaves simple, alternate, elliptic to lance-elliptic, with entire margins that are often rolled under. Flowering October–February; flowers white; fruit an achene.

Infrequent on rocky hillsides or caliche cuestas in the southern Rio Grande Plains (Duval, Hidalgo, Jim Hogg, Starr, and Zapata counties).

BROOM SNAKEWEED, BROOMWEED, IRONWEED
Gutierrezia sarothrae (Pursh) Britt. & Rusby

Erect, much-branched shrublet 15–90 cm tall. Leaves alternate, linear, entire, and resinous. Flowering summer–fall; flowers yellow; fruit an achene.

Locally abundant in calcareous soils in the Trans-Pecos, High and Rolling Plains, and Rio Grande Plains (Duval, Jim Wells, and Live Oak counties).

The foliage is poisonous to livestock.

TATALENCHO

Gymnosperma glutinosum (Spreng.) Less.

Shrubby plant 5–20 dm tall, herbaceous above and woody near the base with slender, glutinous stems and leaves. Leaves simple, alternate, linear with entire margins. Flowering summer–fall; flowers yellow; fruit an achene.

Found on rocky soils of dry hillsides and arid grasslands in the Trans-Pecos, Edwards Plateau, and Rio Grande Plains.

COMMON GOLDENWEED

Isocoma coronopifolia (Gray) Greene

Bushy half-shrub usually 30–90 cm tall with numerous upright herbaceous branches arising from a perennial, woody base. Leaves alternate, thin and linear, resinous, usually pinnately lobed. Flowering September–November; flowers yellowish gold, very showy; fruit an achene.

Frequent in the western Rio Grande Plains on dry calcareous soils.

This plant is nonpalatable to grazing animals and its presence indicates overgrazing.

DRUMMOND GOLDENWEED

Isocoma drummondii (T. & G.) Greene

Bushy half-shrub usually 30–90 cm tall, herbaceous above, woody toward the base with resinous leaves. Leaves simple, alternate, usually linear with mostly entire margins, but rarely the lower leaves have lobed margins. Flowering spring–fall (mostly fall); flowers golden yellow; fruit an achene.

Locally abundant in the Rio Grande Plains (more prevalent in Coastal Prairies and Marshes of south Texas) (Brooks, Cameron, Jim Wells, Kenedy, Kleberg, Nueces, San Patricio, and Willacy counties).

The leaves of this plant are much broader than those of common goldenweed and are usually entire, and the flower heads are much larger.

MARIOLA

Parthenium incanum H. B. K.

Intricately branched shrub 4–10 dm tall with branches and leaves covered with cottony pubescence. Leaves lyrate, broadly oblong to obovate, lobed, 1–6 cm long. Flowering summer–fall; flowers cream; fruit an achene.

In calcareous soils in the Trans-Pecos, High and Rolling Plains, Edwards Plateau, and along the Rio Grande in the Rio Grande Plains (Starr County).

RIDDELL GROUNDSEL

Senecio spartioides T. & G.

Erect, freely branched subshrub usually 5–10 dm tall. Leaves alternate, pinnatifid with linear lobes, 3–6 cm long, often slightly succulent or fleshy. Flowering July–October; flowers yellow; fruit an achene.

Locally abundant in loose, sandy soil in the western half of Texas and the coastal part of the Rio Grande Plains (Brooks, Hidalgo, Jim Hogg, Kenedy, and Willacy counties).

It is poisonous to cattle, horses, sheep, and goats.

MEXICAN TRIXIS

Trixis inula Crantz

[syn. *T. radialis* (L.) O. Ktze.]

Much-branched shrub 1–2 m tall with bright yellow flowers. Leaves simple, alternate, elliptic to lanceolate or oval, with slightly toothed margins. Flowering July–March; flowers yellow, fruit an achene. Frequent in the southern Rio Grande Plains (Cameron, Hidalgo, Jim Hogg, Kenedy, Starr, and Willacy counties) in brush on well-drained, clay loam soils.

White-tailed deer occasionally browse the leaves of this species.

SALADILLO

Varilla texana Gray

Much-branched subshrub 2–3 dm tall, much-spreading and forming clumps. Leaves opposite or nearly alternate, linear, very thick and succulent with entire margins. The leaves are confined to the lower part of the stem. Flowering April–July, often again September–October; flowers yellow; fruit an achene.

Locally abundant in gypseous or saline soil of the Rio Grande Plains (Cameron, Hidalgo, Starr, Webb, and Zapata counties).

RESIN-BUSH, SKELETON-LEAF GOLDENEYE

Viguiera stenoloba Blake

Much-branched shrub with a rounded top attaining a height of 1 m. Leaves alternate to opposite, sometimes linear but usually lobed with entire or partially toothed margins. Flowering in summer; flowers yellow; fruit an achene.

Found on rocky soils in the Trans-Pecos, western Edwards Plateau, and Rio Grande Plains.

It is occasionally browsed by livestock in times of nutritional stress.

ORANGE ZEXMENIA

Wedelia hispida H.B.K.

[syn. *Zexmenia hispida* (H. B. K.) Gray]

Perennial shrub 5–10 dm tall with slender, branched stems,and leaves very rough to the touch. Leaves simple, opposite or alternate, rhombic-lanceolate or ovate-lanceolate, with sparingly toothed or lobed margins. Flowering summer–fall; flowers yellow; fruit an achene.

On dry soil of hillsides and arroyo banks in the Trans-Pecos, Edwards Plateau, Coastal Prairies and Marshes, and Rio Grande Plains.

White-tailed deer occasionally browse the leaves of this plant.

SHORTHORN ZEXMENIA

Zexmenia brevifolia Gray

Perennial herb or rounded shrub 5–10 dm tall, older specimens becoming quite woody and much-branched near the base. Leaves alternate or opposite, ovate to broadly elliptic-ovate, with entire or toothed margins. The leaf surfaces are very rough. Flowering summer–fall; flowers yellow; fruit an achene.

Frequent in low, shrubby vegetation on calcareous or limestone soils of hillsides in the western Rio Grande Plains and the southern Trans-Pecos.

ZINNIA

Zinnia acerosa (DC.) Gray

Small, perennial shrublet attaining a maximum height of 16 cm. This plant is characterized by having a woody base and taproot. Leaves opposite, linear, and entire. Flowering summer–fall; flowers white; fruit an achene.

This is a frequent species in the Trans-Pecos area of western Texas but is rare in the Rio Grande Plains where it is found on rocky outcrops in Starr and Jim Hogg counties.

CONVOLVULACEAE

TREE MORNING GLORY

Ipomoea aristolochiifolia (H.B.K.) G. Don ssp. *fistulosa*

[syn. *I. fistulosa* Mart.]

Subshrubby plant with erect branches attaining a height of 2 m. Leaves very large, simple, alternate, ovate-lanceolate with entire margins. Flowering throughout the year; flowers pink and showy; fruit a small capsule.

A native of South America, but introduced by the Spanish in the vicinity of Brownsville, Texas, and now growing wild throughout the extreme southern Rio Grande Plains (lower Rio Grande Valley). Usually found in waste places or on the edges of resacas.

CUPRESSACEAE

MONTEZUMA BALD CYPRESS, SABINO, AHUEHUETE
Taxodium mucronatum Ten.

Large tree with a straight trunk and enlarged base, reaching a height of up to 30 m. Roots large and spreading. Leaves alternate, linear, straight, or somewhat incurved. Flowering in February; male and female flowers separate on the same plant; fruit a cone.

Infrequent to rare along the lower Rio Grande in the extreme southern Rio Grande Plains (Cameron, Hidalgo, and Starr counties). At one time this species was more common but now it is rare in the U.S. This plant is widespread in Mexico.

EBENACEAE

CHAPOTE, MEXICAN PERSIMMON, TEXAS PERSIMMON, BLACK PERSIMMON
Diospyros texana Scheele

A smooth-barked shrub or small tree rarely 16 m tall. Leaves alternate, broadly ovate to oblong-obovate, with entire margins. The leaves are dark green above, grayish green below, and covered with pubescence. Flowering February–June; flowers white; fruit a three- to eight-seeded, black berry. The male and female flowers are on separate plants.

Found throughout the western two-thirds of Texas in rocky, open woodlands, open slopes, and arroyos. This is a frequent species in the Rio Grande Plains and southern Coastal Prairie.

The fruit is eaten by several species of birds and mammals including the Rio Grande turkey, coyote, javelina, and raccoon. Both the leaves and fruits are eaten by white-tailed deer.

EPHEDRACEAE

CLAPWEED, POPOTE, CANATILLA
Ephedra antisyphilitica C. A. Mey.

Erect or spreading shrub attaining a height of 1 m with green, yellow green, or gray green stems. Leaves opposite, very small, and connate. The leaves are so small and inconspicuous that this plant appears leafless. Flowering winter–spring; male and female flowers on separate plants; fruit a small, reddish cone.

Found on gravelly or rocky hillsides, calcareous slopes, arroyos, ravines, and canyons in central and western Texas and south through the Rio Grande Plains.

It is often heavily browsed by livestock and white-tailed deer. The young tender joints contain about 12 percent crude protein.

EUPHORBIACEAE

VASEY ADELIA
Adelia vaseyi (Coulter) Pax and Hoffm.

Erect, usually narrow shrub 1–3 m tall with many long, slender, upright stems from the base. Leaves scant, alternate or clustered at the nodes, spatulate or linear-spatulate, with entire margins. Flowering January–June; flowers white, male and female flowers on separate plants; fruit a three-lobed, three-seeded capsule.

Infrequent in chaparral on loamy soils in the extreme southern Rio Grande Plains (Cameron and Hidalgo counties).

SOUTHWEST BERNARDIA, OREJA DE RATÓN

Bernardia myricaefolia (Scheele) Wats.

Densely branched, unarmed shrub usually 10–25 dm tall. Leaves simple, alternate or clustered, narrowly elliptic or variable in shape, with toothed margins. The leaves are covered with pubescence. Flowering spring–fall; male and female flowers separate on the same plant or on different plants; fruit a three-lobed, three-seeded capsule.

Infrequent on dry, rocky slopes in the Edwards Plateau and Rio Grande Plains.

This plant is drought resistant and is browsed by cattle in time of nutritional stress. The seeds are eaten by bobwhite quail.

PALILLO, CORTES CROTON

Croton cortesianus H.B.K.

Shrub 1–2 m tall with branches usually 2–3 branched and the younger parts covered with pubescence. Leaves simple, alternate, deciduous, oblong to lanceolate, with entire or toothed margins. Flowering spring–fall; male and female flowers on separate plants; fruit a round or shallowly lobed capsule.

Infrequent in thickets in the extreme southern Rio Grande Plains (Cameron, Hidalgo, and Starr counties). Also present on the Welder Wildlife Refuge in San Patricio County.

ROSVAL, HIERBA DEL GATO, RUBALDO

Croton dioicus Cav.

Strong-scented, slender, erect perennial herb 3–6 dm tall. This plant is silvery throughout as a result of the dense covering of scales. Leaves simple, alternate, deciduous, narrowly oblong to lanceolate or elliptic, with entire or wavy margins. Flowering spring–summer; male and female flowers on separate plants, rarely on same plant; fruit a three-lobed capsule.

Found in the western Rio Grande Plains and throughout the Trans-Pecos.

SALVIA, BERLANDIER CROTON

Croton humilis L.

Strong-scented, slender, weak shrub 5–23 dm tall usually growing up through other shrubs. Leaves simple, alternate, deciduous, ovate to oblong or lanceolate, with entire or wavy margins. Flowering nearly all year following rains; flowers white, male and female flowers on separate or same plant; fruit a three-lobed capsule.

Locally abundant in chaparral in the Rio Grande Plains (Cameron, Hidalgo, Nueces, Starr, Willacy, and Zapata counties).

HIERBA DEL JABALI, BEACH-TEA

Croton punctatus Jacq.

Strong, perennial herb or small shrub attaining a height of 30 cm. This plant spreads by underground rhizomes and grows in rounded mounds. Leaves alternate, oval to oblong or ovate to elliptic, with undulated or entire margins. The leaves are covered with fine pubescence. Flowering throughout the year; male and female flowers on separate plants, rarely on the same plant; fruit a three-lobed (rarely two- or four-lobed) capsule.

Found in loose, deep sands along the Coastal Prairies and Marshes including the barrier islands in south and southeast Texas.

TORREY CROTON

Croton incanus H.B.K.

[syn. *C. torreyanus* Muell. Arg.]

A whitish, slender shrub 1–2 m tall. Branches slender and straight, bearing minute hairs. Leaves pubescent, simple, alternate, oblong with entire margins. Flowering summer–fall; flowers light green, male and female flowers separate on the same plant; fruit a three-lobed, three-seeded capsule.

Found on rocky ravines and hillsides in dry, gravelly soils in the Trans-Pecos and Rio Grande Plains.

CANDELILLA, WAX EUPHORBIA

Euphorbia antisyphilitica Zucc.

Small, erect shrub attaining a maximum height of 10 dm. It is characterized by having rodlike, leafless stems. This plant has leaves but they are very small and early deciduous, thus seldom seen. Flowering throughout the year following rains; flowers white or pinkish; fruit a three-seeded, three-lobed capsule.

Primarily a Chihuahuan Desert species, it is found in the Trans-Pecos counties of west Texas with isolated populations in Starr and Webb counties of the Rio Grande Plains.

The branches yield a good grade of wax when boiled. For this reason, the plant is of commercial importance in Mexico.

RUBBER-PLANT, LEATHER STEM, SANGRE DE DRAGO

Jatropha dioica Sesse ex Cerv.

Plant forming colonies from underground runners, usually 2–6 dm tall. Stems succulent (scarcely woody), very flexible, and tough. Leaves simple, alternate or fascicled on short branches, spatulate or linear, with entire margins. Flowering July–August; flowers white, male and female flowers on separate plants; fruit a two- or three-seeded capsule.

Found on dry slopes, mesas, and rocky limestone bluffs in the Rio Grande Plains and western Edwards Plateau.

This plant is poisonous to sheep and goats.

FAGACEAE

POST OAK

Quercus stellata Wang.

Moderate-sized tree attaining a height of nearly 25 m with stout limbs and a dense, rounded crown. Leaves simple, alternate, deciduous, oblong-obovate (two- to four-lobed on each side) with wavy margins. Flowering in spring; flowers yellowish, male and female flowers separate on the same plant; fruit an acorn.

Found on sandy soil in dry upland woods in the Pineywoods and Edwards Plateau and west along the Coastal Prairie to San Patricio County.

The acorns are eaten by white-tailed deer, javelinas, fox squirrels, and Rio Grande turkey.

LIVE OAK, ENCINO

Quercus virginiana Mill.

Small or large trees attaining a height of nearly 20 m with a wide-spreading crown and large limbs close to the ground. Leaves simple, alternate, variable in size and shape with entire, toothed, or even lobed margins. The upper leaf surfaces are glossy, whereas the lower surfaces are pubescent. Flowering in spring; flowers yellow, male and female flowers are separate on the same plant; fruit an acorn.

This tree is usually found in sandy or sandy loam soils of the Rio Grande Plains, Coastal Prairie, and Pineywoods.

The acorns are frequently eaten by white-tailed deer, javelinas, fox squirrels, and Rio Grande turkeys.

FLACOURTIACEAE

BRUSH-HOLLY, CORONILLA, MEXICAN XYLOSMA
Xylosma flexuosa (H.B.K.) O. Ktze.

Slender, evergreen, thorny shrub or small tree usually 2–3 m tall. Leaves simple, alternate, ovate-elliptic to elliptic or oblanceolate to obovate, with coarsely and remotely toothed margins. Flowering throughout the year; flowers very small and white; fruit a small, two- to eight-seeded, red berry. Male and female flowers are sometimes on separate plants.

Infrequent to rare in chaparral and palm groves in the extreme southern Rio Grande Plains and northeast along the Coastal Prairies and Marshes to Nueces County.

FRANKENIACEAE

FRANKENIA
Frankenia johnstonii Correll

Low, erect shrub attaining a height of 3 dm with a woody base giving rise to several arching or recurved willowy stems. The entire plant is grayish or bluish green in appearence. Leaves opposite, oblanceolate to oblong-elliptic, with entire margins that are rolled under. Flowering November–April; flowers white; fruit a small capsule.

Rare on saline flats and rocky, gypseous hillsides in the extreme southern Rio Grande Plains (Starr County).

JUGLANDACEAE

PECAN, NOGAL MORADO, NUEZ ENCARCELADA
Carya illinoinensis (Wang.) K. Koch

Large tree with a rounded crown, attaining a height of 50 m. The trunk is often very large and may reach a diameter of 2 m. Leaves alternate, deciduous, odd-pinnately compound with nine- to 17-leaflets (leaflets oblong-lanceolate to lanceolate, with toothed margins). Flowering in spring; flowers yellowish, male and female flowers separate on the same plant; fruit a nut.

Found on rich bottomland soils in the eastern half of Texas (San Patricio County).

It is widely cultivated for the commercially important pecans. These are an important food item to many kinds of wildlife. The pecan is the state tree of Texas.

KOEBERLINACEAE

ALLTHORN, CRUCIFIXION THORN, JUNCO
Koeberlinia spinosa Zucc.

Much-branched, usually leafless shrub or tree attaining a height of 7 m and consisting of a tangled mass of stiff, green spines. Flowering March–October; flowers greenish white; fruit a one- to four-seeded, black berry.

Found on rocky open slopes, clay mounds, brushlands, and about arroyos in the southern Rio Grande Plains and the Trans-Pecos.

This plant is a good example of adaptation to desert conditions with the green thorns and twigs carrying on the photosynthetic process.

KRAMERIACEAE

CALDERONA, RATANY

Krameria ramosissima (Gray) Wats.

Diffusely branched and rigid shrub 3–10 dm tall, all parts except the older branchlets covered with dense, grayish hairs. Leaves very small, simple, alternate or fascicled in the axils, linear to linear-lanceolate, with entire margins. Flowering spring–fall; flowers maroon purple; fruit a nearly globose, one-seeded pod.

Locally abundant on sunny hillsides and in dry, shrubby vegetation in the Trans-Pecos and Rio Grande Plains.

This plant is occasionally browsed by white-tailed deer.

LABIATAE

HORSEMINT

Monarda fruticulosa Epl.

Diffusely branched, aromatic shrub attaining a height of 6 dm with pubescent stems. Leaves opposite, linear with entire margins (occasionally toothed above the middle). Flowering throughout the year; flowers white or pinkish.

In deep sand or sandy loam in prairies, mesquite plains, open woods and fields, and on active dunes in the Rio Grande Plains.

BLUE SALVIA, BLUE SAGE, MEJORANA, CRESPA
Salvia ballotaeflora Benth.

Aromatic, much-branched, square-stemmed shrub attaining a height of 25 dm. Leaves pubescent, simple, opposite with toothed margins. Flowering January–October; flowers bluish or purple; fruit a small nutlet.

Found on rocky, sandy, or gravelly soils on brushy slopes, thickets, and chaparral in the Rio Grande Plains.

The dried leaves are used for flavoring meats and other foods.

LAURACEAE

RED BAY, SWEETBAY
Persea borbonia (L.) Spreng.

Evergreen shrub or tree to 20 m tall. Leaves simple, alternate, lanceolate to elliptic or elliptic-oblanceolate, with entire margins. Flowering May–June; flowers pale yellow; fruit a dark blue or blackish drupe.

Found along Coastal Prairies and Marshes of southeast Texas (southwest to San Patricio and Nueces counties).

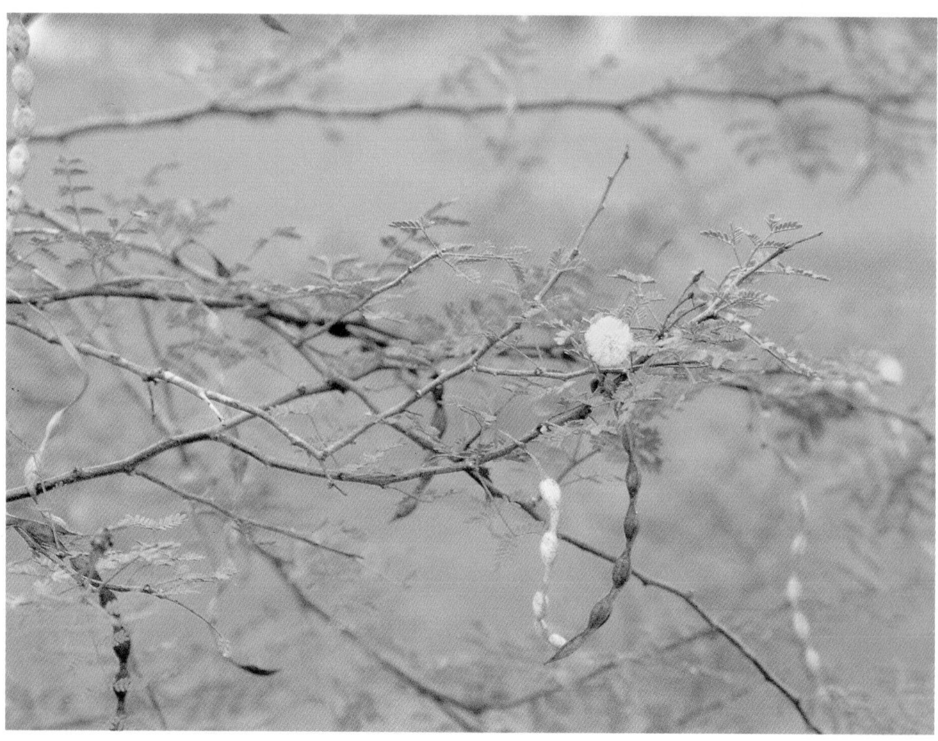

LEGUMINOSAE

GUAJILLO
Acacia berlandieri Benth.

Spreading shrub with many stems from the base or sometimes a small tree 1–2 m tall (rarely 4 m). The gray to white branches are usually armed with scattered, nearly straight prickles. Leaves delicate and almost fernlike in appearance, alternate, bipinnately compound with five to nine pairs of pinnae (30–50 pairs of leaflets per pinnae; leaflets linear to oblong). Flowering November–March; flowers white; fruit a five- to ten-seeded legume.

Abundant on limestone and caliche cuestas of the Rio Grande Plains, Trans-Pecos, and southern Edwards Plateau.

This plant is browsed by cattle and white-tailed deer but is poisonous to sheep and goats when eaten for prolonged periods. The leaves have a crude protein content of about 20 percent. Guajillo is an important source of nectar for honey production.

MESCAT ACACIA
Acacia constricta Gray

Spiny shrub 1–3 m tall with slender, paired thorns at the nodes. Leaves alternate, bipinnately compound with four to seven pairs of pinnae (six to 16 pairs of leaflets per pinnae; leaflets oblong to linear). Flowering in spring or later following rains in droughty years; flowers yellow; fruit a reddish brown to black legume that is much-constricted between the seeds.

An abundant species in the Trans-Pecos and southwestern High and Rolling Plains but rare in the southwestern Rio Grande Plains (Starr and Zapata counties).

HUISACHE

Acacia farnesiana (L.) Willd.

Shrub or usually small tree 2–4 m tall, often with several trunks flaring upward. Leaves alternate, bipinnately compound with two to eight pairs of pinnae (10 to 25 pairs of leaflets per pinnae; leaflets linear-oblong). Flowering February–March; flowers yellowish gold; fruit a reddish brown to purple or black legume with two rows of seeds.

Found throughout the Rio Grande Plains and southern Coastal Prairie, north to the southern Edwards Plateau, and west to the Trans-Pecos.

It is considered a good winter forage plant for cattle and the pollen produced by the flower provides food for bees. White-tailed deer are known to browse the leaves, stems, and fruit. Javelina occasionally eat the fruit. It is an important nesting species for mourning doves. The plants may form dense stands on pasture and rangeland that require control measures to reduce densities.

CATCLAW, UÑA DE GATO

Acacia greggii Gray

Thorny, thicket-forming shrub usually not more than 1–2 m tall, rounded and much branched. Leaves alternate, bipinnately compound with one to three pairs of pinnae (three to seven pairs of leaflets per pinnae; leaflets obovate to narrowly oblong). Flowering April–October; flowers creamy yellow; fruit a light brown to reddish legume often becoming constricted between the seeds.

Frequent in chaparral on the Rio Grande Plains, Trans-Pecos, and Edwards Plateau; infrequent in the southern Coastal Prairie.

This plant often grows in almost impenetrable thickets, furnishing shelter for wildlife. The seeds are eaten by bobwhite and scaled quail; white-tailed deer and cattle browse the foliage; the pollen produced by the flowers is an important food for bees and the nectar is important for honey production. The leaves contain 16–20 percent crude protein.

BLACKBRUSH ACACIA, CHAPARRO PRIETO

Acacia rigidula Benth.

Thorny shrub or small tree with many stems forming the base, attaining a height of 1–3 m. Leaves alternate, bipinnately compound with one or two pairs of pinnae (usually one) and two to four pairs of leaflets (rarely five) per pinnae (leaflets elliptic to oblong). Flowering February–July; flowers white or light yellow; fruit a reddish brown to black legume.

Abundant on sandy loam and calcareous soils in the Rio Grande Plains, southern Coastal Prairie, and Trans-Pecos where it often forms impenetrable thickets.

The seeds are eaten by bobwhite quail. The leaves and beans are browsed by white-tailed deer; the flowers are a source of nectar for honey. The leaves contain about 15 percent crude protein.

ROEMER ACACIA

Acacia roemeriana Scheele

Round-topped, spiny shrub 1–3 m tall with many spreading branches. Leaves alternate, bipinnately compound with one to three pairs pinnae, and four to eight pairs of leaflets per pinnae (leaflets obtuse, strongly veined). Flowering in spring or later following rains in droughty years; flowers creamy white; fruit a brown legume.

Frequent in chaparral in southern Trans-Pecos, infrequent in Edwards Plateau and northern Rio Grande Plains, and rare in southern Rio Grande Plains (Starr County).

TWISTED ACACIA, HUISACHILLO

Acacia schaffneri (Wats.) Herm.

[syn. *A. tortuosa* L. Willd.]

Usually a spiny, spreading shrub 5–15 dm (rarely 20 dm) tall, very similar to huisache but differing by the spreading growth form and the narrower and longer legume. Leaves alternate, bipinnately compound with two to five pairs of pinnae (10 to 15 pairs of leaflets per pinnae; leaflets linear-oblong). Flowering in spring and later following rains in droughty years; flowers yellow; fruit a dark reddish brown to black legume.

Locally frequent in chaparral in the Rio Grande Plains and southern Coastal Prairie.

The leaves and stems are browsed by white-tailed deer.

PRAIRIE ACACIA

Acacia texensis T. & G.

Rounded subshrub 1–5 dm tall often forming colonies by means of woody rhizomes. Leaves alternate, bipinnately compound with three to eight pinnae and six to 20 pairs of leaflets per pinnae (leaflets linear, glabrous, or appressed pubescent). Flowering May–August; flowers whitish cream; fruit a brownish legume.

Locally abundant in open chaparral in the Trans-Pecos, Edwards Plateau, and Rio Grande Plains (Hidalgo and Starr counties).

WRIGHT ACACIA, UÑA DE GATO

Acacia wrightii Benth.

Spiny shrub or sometimes a small tree attaining a height of 2–3 m. The spreading branches form a wide, irregularly shaped crown. Leaves solitary or fascicled, bipinnately compound with one to three pairs of pinnae (two to six pairs of leaflets per pinnae; leaflets obovate to narrowly oblong). Flowering March–May or at odd times after rains; flowers creamy yellow; fruit a broad, light brown legume.

Locally frequent in chaparral and woodlands along creeks and canyons in the Rio Grande Plains, Edwards Plateau, and Trans-Pecos.

This plant is closely related to catclaw but can be distinguished by the wider legume and larger leaflets. The pollen produced by the flowers is an important food for bees and a source of nectar for honey.

CAESALPINIA

Caesalpinia caudata (Gray) Fisher

Low, branched shrub 3–5 dm high. Leaves alternate, bipinnately compound with three to 11 pairs of pinnae and three to 11 pairs of leaflets per pinnae (leaflets broadly ovate). Flowering in spring; flowers yellow; fruit a small two- to four-seeded legume.

Found in loose, sandy soils of the Rio Grande Plains.

FALSE-MESQUITE CALLIANDRA

Calliandra conferta Gray

Low, densely branched shrub usually 1–3 dm tall. This plant resembles some of the *Mimosa* species (pp. 109–111) but is separated from them by its lack of spines. Leaves bipinnately compound with one pair of pinnae (rarely two pairs) and a few pairs of small leaflets (leaflets ovate to oblong). Flowering March–May; flowers reddish purple; fruit a small, gray to black legume.

Locally abundant on caliche and limestone soils in the Trans-Pecos, Edwards Plateau, and Rio Grande Plains.

Most plants are closely hedged by cattle and white-tailed deer.

LINDHEIMER SENNA

Cassia lindheimeriana Scheele

Erect, velvety-hairy, perennial herb becoming bushy in appearance and attaining a height of 1–2 m. The base of this plant is woody. Leaves spirally arranged, densely pubescent, and pinnately compound with five to eight pairs of leaflets (leaflets oblong to elliptic). Flowering August–October; flowers yellow; fruit a broadly linear, many-seeded legume.

Usually found on limestone soils of the Trans-Pecos, Edwards Plateau, and Rio Grande Plains.

TEXAS BABY-BONNETS

Coursetia axillaris Coult. & Rose

Densely branched shrub or small tree 5–15 dm tall having no thorns. Leaves alternate, pinnately compound with three to five pairs of leaflets (leaflets broadly elliptic or obovate). Flowering spring–fall; flowers pale pink; fruit a two- to six-seeded reddish brown legume.

Rare in scrubby chaparral on caliche ridges and in woodlands of the Rio Grande Plains and southern Coastal Prairie (Duval, Hidalgo, Jim Hogg, and San Patricio counties).

THYRSUS DALEA

Dalea scandens (P. Mill.) Clausen var. *pauciflora* (Coult.) Barneby

[syn. *D. thyrsiflora* Gray]

Small, erect subshrub 5–12 dm tall with herbaceous branches and woody stems. The branches and leaves are covered with pubescence. Leaves alternate, pinnately compound with three to five leaflets (leaflets obovate or ovate, with entire margins). Flowering September–December; flowers yellow to brownish purple; fruit a very small legume usually enclosed by a calyx.

Found on clay or sandy loam soil, often along arroyos in the extreme southern Rio Grande Plains (Cameron County).

CORAL BEAN, COLORIN

Erythrina herbacea L.

Thorny shrub or subshrub with slender spreading stems from the base, attaining a height of 20 dm. Leaves alternate, trifoliate, hastately deltoid with entire margins. Flowering April–June; flowers scarlet red; fruit a black legume containing scarlet red beans.

Found on sandy soils along Coastal Prairies and Marshes of south and southeast Texas and inland in the east Texas Pineywoods.

It is frequently grown as an ornamental. The beans are poisonous.

TEXAS KIDNEYWOOD, VARA DULCE

Eysenhardtia texana Scheele

Irregularly shaped, spineless shrub (rarely a small tree) 2–3 m tall. The vegetative parts of this plant have an unpleasant odor when bruised. Leaves alternate, pinnately compound with 15 to 47 leaflets per leaf (leaflets oblong). Flowering April–September, frequently after rains; flowers white; fruit a very small, green to brown legume.

Frequent in chaparral vegetation on calcareous soils of the Trans-Pecos, Edwards Plateau, southern Coastal Prairie, and Rio Grande Plains.

The leaves are browsed by white-tailed deer.

INDIGO
Indigofera suffruticosa Mill.

Erect, perennial plant 5–20 dm tall, herbaceous above but woody near the base. Leaves pinnately compound, alternate; leaflets oblong to elliptic or oblaceolate to obovate, with entire margins. Flowering July–November; flowers reddish; fruit a small legume.

Found locally in Coastal Prairies and Marshes of south (Cameron and San Patricio counties) and southeast Texas, also inland to the south-central Edwards Plateau and parts of the east Texas Pineywoods.

This plant was one of the sources of indigo, the blue dye formerly important in commerce.

POPINAC, WHITE POPINAC LEAD-TREE
Leucaena leucocephala (Lam.) de Wit.

Tree 2–8 m tall with numerous, unarmed, spreading branches. Leaves alternate, bipinnately compound with four to nine pairs of pinnae and 11 to 17 pairs of leaflets per pinnae (leaflets oblong to laceolate, with entire margins). Flowering May–June; flowers creamy white; fruit a strap-shaped legume.

Scattered along streams and resacas in the extreme southern Rio Grande Plains (Cameron and Hidalgo counties).

This tree was introduced and planted as an ornamental but has escaped cultivation and is now naturalized.

TEPEGUAJE, LEAD TREE

Leucaena pulverulenta (Schlecht.) Benth.

Tree 3–10 m tall (rarely 13 m) with smooth, gray to brown branches spreading into a broad, rounded crown. Leaves alternate, bipinnately compound with 28 to 36 pairs of pinnae and 15 to 60 pairs of leaflets per pinnae (leaflets linear). Flowering March–June; flowers white; fruit a strap-shaped legume with the seeds transversely arranged.

Abundant along streams and resacas in the extreme southern Rio Grande Plains (Cameron and Hidalgo counties).

This species is often planted as a yard and street tree. Several species of birds (e.g., grackles) use this tree for cover and nesting sites.

VINE MIMOSA, RASPILLA

Mimosa malacophylla Gray

Vinelike plant climbing in trees or forming a tangle usually 3–4 m high, less commonly a weak-stemmed shrub, the stems armed with recurved prickles. Leaves bipinnately compound with three to seven pairs of pinnae having three to eight pairs of leaflets per pinnae (leaflets ovate to oblong or obovate). Flowering June–July, occasionally later following rains; flowers white; fruit a six- to eight-seeded legume.

Rare in woodlands in the extreme southern Rio Grande Plains.

It is often used as a nesting site by birds.

ZARZA, COATANTE

Mimosa pigra L.

Shrub 1–3 m tall, much-branched, the branches armed with stiff, slightly recurved prickles. Leaves alternate, bipinnately compound with four to six (occasionally as many as 12) pairs of pinnae and 20 or more pairs of leaflets per pinnae (leaflets linear to oblong). Flowering March–November; flowers pink; fruit an eight- to 15-seeded legume.

Locally abundant in dry lake beds and resacas and other seasonally inundated areas of clay soil in the extreme southern Rio Grande Plains.

WHERRY MIMOSA

Mimosa wherryana (Britt. & Rose) Standl.

Rounded, much-branched, prickly shrub 5–15 dm (rarely 20 dm) tall. Leaves bipinnately compound with one to three pairs of pinnae and three to six pairs of leaflets per pinnae (leaflets oblong). Flowering spring–fall; flowers white; fruit a small legume with sharp, slender prickles on the margins.

On caliche and gravelly hills near the Rio Grande in Starr and Zapata counties in the extreme southwestern Rio Grande Plains.

RETAMA

Parkinsonia aculeata L.

Green-barked, thorny shrub or tree to 10 m tall with slender, spreading branches often forming a rounded crown. Leaves alternate, bipinnately or rarely pinnately compound with one or two pairs of pinnae and numerous pairs of leaflets per pinnae (leaflets linear to oblanceolate). Flowering spring–fall; flowers yellow; fruit a many-seeded, brown legume.

Frequent in the Rio Grande Plains and southern Coastal Prairie, especially in low, poorly drained areas.

This woody plant is occasionally browsed by white-tailed deer. The seeds are eaten by bobwhite quail.

TEXAS PALOVERDE, PALOVERDE

Parkinsonia texana (Gray) Wats.

[syn. *Cercedium texanum* Gray]

Spiny, green-barked shrub or small tree usually less than 2 m tall. This plant is usually several-stemmed from the base and early deciduous, leaving bare twigs and branches. Leaves alternate, bipinnately compound with usually only one pair of pinnae (rarely two pairs) and one to three pairs of leaflets per pinnae (leaflets oblong). Flowering spring–summer; flowers yellow; fruit a one- to four-seeded, dark brown legume.

Locally abundant in semidesert scrub in the western half of the Rio Grande Plains.

White-tailed deer occasionally browse this woody plant. The seeds are eaten by several species of birds and rodents.

BORDER PALOVERDE, PALOVERDE

Parkinsonia texana (Gray) Wats. var. *macrum* (I. M. Johns.) Isely

[syn. *Cercidium macrum* I. M. Johnst.]

Green-barked, spiny tree attaining a height of 3–4 m. Leaves early deciduous or releafing after rains, alternate, bipinnately compound with two to three pairs of pinnae and two to three pairs of leaflets per pinnae (leaflets oblong). Flowering spring–fall; flowers yellow; fruit a one- to five-seeded, dark brown legume.

On sandy loam or clay soils of the Rio Grande Plains inland to McMullen, Duval, and Starr counties.

Border paloverde is closely related to Texas paloverde and separation often depends on legume and flower characteristics.

TEXAS EBONY, EBANO

Pithecellobium ebano (Berlandier) C.H. Muller

[syn. *P. flexicaule* (Benth.) Coult.]

Spiny, evergreen shrub or small tree sometimes attaining a height of 15 m and forming a rounded crown. Leaves alternate, bipinnately compound with one to three pairs of pinnae and three to six pairs of leaflets per pinnae (leaflets obliquely elliptic, oval, or obovate). Flowering April–July, rarely to November; flowers yellowish or cream; fruit a large, many-seeded, dark brown or black legume.

Frequent in the southern Rio Grande Plains.

Prior to clearing of agricultural lands beginning in the 1920s, this tree was very abundant and used as a preferred nest site by white-winged doves. It is often planted as an ornamental or shade tree. White-tailed deer are known to browse the leaves, and the seeds are eaten by small rodents and javelina.

TENAZA, HUAJILLO

Pithecellobium pallens (Benth.) Standl.

Spiny shrub usually 1–2 m tall, rarely a small tree to 6 m. Leaves bipinnately compound with three to six pairs of pinnae and seven to 20 pairs of leaflets per pinnae (leaflets oblong-linear). Flowering May–August; flowers white; fruit a many-seeded, reddish brown legume.

Locally abundant in chaparral on alluvial soils of stream bottoms or on edges of water holes in the Coastal Prairies and Marshes of south Texas north to San Patricio County and in the extreme southern Rio Grande Plains (Cameron, Hidalgo, Starr, and Willacy counties).

MESQUITE, HONEY MESQUITE

Prosopis glandulosa Torr.

Thorny shrub or small tree attaining a height of 10 m. Leaves alternate, bipinnately compound with one to several pairs of pinnae and six to 20 pairs of leaflets per pinnae (leaflets linear). Flowering May–September; flowers yellowish green; fruit a many-seeded, shiny brown legume.

This plant is scattered throughout Texas but is most prevalent in the western half of the state.

The pods (legumes) are an important source of food for horses, cattle, goats, white-tailed deer, and javelina. They are mildly poisonous to cattle and goats if eaten over prolonged periods. Mesquite provides nesting sites for several species of birds including the white-winged dove, mourning dove, and chachalaca. The flowers are an important source of nectar for honey production. The plants frequently form dense thickets on pasture and rangeland, often requiring control measures to reduce densities.

TORNILLO, CREEPING MESQUITE, DWARF SCREW-BEAN
Prosopis reptans Benth.

Low undershrub usually 2–4 dm tall. Stems slender and wiry with long, gray thorns. Leaves alternate, bipinnately compound with one pair of pinnae and seven to 13 pairs of leaflets per pinnae (leaflets oblong). Flowering spring–fall; flowers dark yellow; fruit a yellow to reddish brown legume that is tightly spirally coiled.

Found on clay soils in the Rio Grande Plains, especially common along Coastal Prairies of south Texas.

This plant is usually low and easily overlooked. The oddly twisted pod is a good mark of identification. White-tailed deer browse the leaves.

RATTLEBUSH, SIENE BEAN, COFFEE BEAN
Sesbania drummondii (Rydb.) Cory

A short-lived shrub 4–30 dm tall with many stems from the base. Leaves alternate, pinnately compound with numerous pairs of leaflets (leaflets narrowly oblong or elliptic). Flowering June–September; flowers yellow; fruit a four-seeded, four-winged legume.

Found in low, wet places in the east Texas Pineywoods, Coastal Prairies and Marshes, and the Rio Grande Plains.

The seeds are poisonous to sheep, goats, and cattle and may cause death if eaten in large quantities.

MESCAL BEAN, TEXAS MOUNTAIN LAUREL, FRIJOLILLO

Sophora secundiflora (Ort.) Lag. ex DC.

Evergreen shrub or sometimes a small tree 5–35 dm tall usually with dense, dark green, glossy foliage and velvety twigs. Leaves alternate, pinnately compound with five to 13 leaflets per leaf (leaflets elliptic-oblong or oval). Flowering March–April; flowers violet, very showy; fruit a one- to eight-seeded, brown, pubescent legume.

Frequent on caliche or limestone soils in the Trans-Pecos and southern Edwards Plateau; infrequent in the Rio Grande Plains.

The leaves and seeds are poisonous to cattle, sheep, and goats. It is an important ornamental and can be propogated from scarified seed.

YELLOW SOPHORA

Sophora tomentosa L.

Small, rounded shrubs usually about 1 m tall with densely pubescent foliage. Leaves alternate, pinnately compound with 13 to 21 leaflets (leaflets elliptic, oblong, or oval). Flowering March–October; flowers bright yellow; fruit a densely pubescent legume.

Infrequent along the Coastal Prairie in south Texas as far north as Aransas County. The seeds are poisonous.

LILIACEAE

BUCKLEY YUCCA
Yucca constricta Buckl.

Plant usually stemless but sometimes developing a short stem to 4 dm high. Leaves linear with long, curly, filiform threads on the margins. The leaf tips possess a stout, short spine. Flowering April–June; flowers pale, greenish white; fruit a brown or black capsule.

Thinly scattered in chaparral and grasslands of the Rio Grande Plains, Edwards Plateau, and Trans-Pecos.

SPANISH DAGGER, PALMA PITA
Yucca treculeana Carr.

Tree usually 3–3.5 m high with a simple trunk or with a few stout, spreading branches at the top. The crown is comprised of large, symmetrical heads of radiating, sharp-pointed leaves. Flowering February–April; flowers creamy white; fruit a many-seeded, reddish brown or black capsule.

Found in the southern Edwards Plateau and throughout the Rio Grande Plains on well-drained hillsides, chaparral regions, and open flats in the Coastal Prairie near the Gulf of Mexico.

White-tailed deer and cattle browse the leaves, and the trunk is eaten by javelina. Several species of birds use this plant as a nesting site including the Harris' hawk and cactus wren. It is often used as an ornamental. The flowering heads are often the first sign of spring in south Texas and are harvested and eaten as a salad or sautéed with nopalitos and onions.

LOASACEAE

STINGING CEVALLIA, STICK-LEAF CEVALLIA
Cevallia sinuata Lag.

Low-growing, herbaceous perennial to 6 dm tall with a woody base. Leaves and stems armed with long, stinging hairs. Leaves alternate with wavy margins. Flowering June–October; flowers yellow; fruit an achene.

Usually found in open areas, often along roadsides and on various coarse substrates in the western half of Texas. Infrequent near the Rio Grande in the Rio Grande Plains.

LOGANIACEAE

TEPOZAN, BUTTERFLY-BUSH
Buddleja sessiliflora H.B.K.

Shrub to 2 m tall or more with the young stems and leaves covered with short, densely matted, white hair. Leaves opposite, narrowly lanceolate to oblong or ovate, with entire or toothed margins. Flowering spring–summer; flowers greenish yellow; fruit a small capsule.

On sandbars and banks and in palm groves in the extreme southern Rio Grande Plains (Cameron and Hidalgo counties).

LYTHRACEAE

HACHINAL, WILLOW-LEAF HEIMIA
Heimia salicifolia (H.B.K.) Link & Otto.

Spreading, much-branched shrub attaining a height of 3 m, usually much smaller. Leaves simple, opposite or alternate, linear-oblanceolate or lanceolate, with entire margins. Flowering March–November; flowers yellow; fruit a four-celled, brown capsule containing minute seeds.

Usually found along resacas, streams, and in wet soil in the southern Rio Grande Plains.

MALPIGHIACEAE

BARBADOS CHERRY, MEXICAN MYRTLE, MANZANITA
Malpighia glabra L.

Erect shrub attaining a height of 25 dm with many slender stems from the base. Leaves simple, opposite, ovate to elliptic-lanceolate, with entire margins. Flowering March–October; flowers pink; fruit a red, three-celled, three-lobed drupe.

In thickets, brushlands, and palm groves in the southern Coastal Prairie and southern Rio Grande Plains.

The fruit is edible and is sometimes made into preserves. The plant is often cultivated and makes a desirable shrub for gardens. White-tailed deer occasionally browse the leaves, and the fruits are eaten by the coyote and raccoon.

MALVACEAE

PSEUDOABUTILON
Allowissadula lozanii (Rose) Bates
[syn. *Pseudoabutilon lozanii* (Rose) R. E. Fries]

Subshrub attaining a maximum height of about 15 dm with branches and leaves covered with pubescence. Leaves large, alternate, suborbicular-ovate to triangular-ovate, with toothed margins. Flowering throughout the year; flowers yellow; fruit a capsule.

Found on clay soil in thickets and open grassy areas throughout the Rio Grande Plains and the eastern Edwards Plateau.

TULIPAN DEL MONTE, MOUNTAIN ROSE-MALLOW, HEART-LEAF HIBISCUS
Hibiscus cardiophyllus Gray

Woody-based perennial attaining a height of 3–6 dm with densely pubescent branches. Leaves alternate, broadly ovate with toothed margins, and covered with fine pubescence. Flowering throughout the year; flowers very showy, crimson to deep rose red; fruit a several-seeded capsule.

On gravelly hills, about boulders and breaks, and in chaparral in the western Rio Grande Plains and southern Coastal Prairie (Nueces and San Patricio counties).

DRUMMOND WAX-MALLOW, TEXAS MALLOW, WILD TURK'S CAP

Malvaviscus arboreus Cav.

Shrub to 3 m tall covered with short, densely matted, white pubescence. Leaves large, simple, alternate, rounded-cordate, with toothed margins. Flowering throughout the year; flowers bright red; fruit red and berrylike.

Found on slopes and ledges, in wooded arroyos, along streams, and in palm groves in the Rio Grande Plains, southern part of the Edwards Plateau, and along the Coastal Prairies and Marshes of southeast Texas.

The fruit is eaten by several species of passerine birds, and the leaves are occasionally browsed by livestock.

SIDA

Sida filipes Gray

Slender, branched, perennial herb or subshrub attaining a height of 1 m with a woody base. Leaves covered with pubescence, alternate, lanceolate or the lower ones oblong, with coarsely toothed margins. Flowering March–August; flowers deep violet purple; fruit a capsule.

Found in rocky ravines and on dry limestone hills and flats in the Rio Grande Plains and Edwards Plateau.

MELIACEAE

CHINABERRY-TREE, PRIDE-OF-CHINA, CANELÓN, PARAÍSO
Melia azedarach L.

Tree to 15 m tall with a broad spreading crown. Leaves large, deciduous, alternate, bipinnately compound (leaflets numerous, ovate to elliptic-lanceolate, with toothed or entire margins). Flowering March–May; flowers white to pale lavender; fruit a three- to five-seeded, yellow drupe.

Found in thickets, floodplain woods, and borders of woods in the eastern half of Texas (scattered throughout the Rio Grande Plains and southern Coastal Prairie).

This tree was introduced into the United States as an ornamental and is native of Asia. It is widely cultivated but escapes and grows wild. The fruit and leaves are poisonous to goats and pigs. The fruits are consumed by several species of birds.

MORACEAE

RED MULBERRY, MORAL
Morus rubra L.

Tree to 20 m tall with a short trunk and a rather broad, spreading crown. Leaves simple, alternate, deciduous, broadly ovate to ovate-oblong, with toothed margins. Flowering March–May; male and female flowers separate on the same plant; fruit a dark purple, compressed achene that resembles a blackberry.

In upland woods and flood plains mostly in eastern and central Texas but found at scattered localities in the Rio Grande Plains and southern Coastal Prairie where it has escaped from cultivation.

The fruit is eaten by several species of passerine birds and by the raccoon, opossum, and fox squirrel. This tree is often planted as an ornamental. The fruits are edible by humans.

NYCTAGINACEAE

DEVIL'S CLAW, COCKSPUR, GARABATO PRIETO
Pisonia aculeata L.

Densely branched shrub or semierect vine, often with a thick trunk, 1–2 m tall. The branches are stout, elongate, and usually armed with curved, sharp spines. Leaves mostly opposite, sometimes with smaller ones in their axils, elliptic to oval or ovate, with entire margins. Flowering in spring; flowers small, inconspicuous, and creamy-colored with male and female flowers on different plants; frtuit an utricle.

Found in thickets and old resaca beds in the extreme southern Rio Grande Plains (Cameron and Hidalgo counties).

OLEACEAE

DESERT OLIVE, PANALERO, ELBOW-BUSH
Forestiera angustifolia Torr.

Evergreen, dense, stiff, rounded-branched shrub or rarely a small tree usually 1–2 m tall. Leaves simple, opposite or clustered on short knotty spurs, linear with entire margins. Flowering spring–summer; flowers greenish, male and female flowers on separate plants; fruit a one-seeded, black drupe.

Found on dry, well-drained hillsides, along stony hillsides, and in brush of the Rio Grande Plains and Trans-Pecos.

The fruit is eaten by the coyote, raccoon, and scaled quail. This plant is frequently browsed by livestock and white-tailed deer.

MEXICAN ASH, FRESNO, BERLANDIER ASH, RIO GRANDE ASH
Fraxinus berlandieriana A. DC.

Small, round-topped tree rarely over 10 m tall with grayish green to bright green compound leaves having three to five leaflets with toothed margins. The leaflets are lanceolate to elliptic or obovate in shape. Flowering March–August; male and female flowers on separate plants; fruit a samara.

Found along wooded streams and canyons in the Edwards Plateau and Rio Grande Plains.

Several species of birds utilize Mexican ash as nesting sites. It is widely planted as an ornamental in west and southwest Texas.

PALMAE

TEXAS PALM, SABAL PALM, TEXAN PALMETTO,
RIO GRANDE PALMETTO, PALMA DE MICHAROS
Sabal texana (Cook) Becc.

Tall trees attaining a height of 16 m with a large trunk up to 8 dm in diameter. Leaves alternate, fan-shaped, forming a dense, rounded crown. The dead leaves are persistent on the trunk. It is very similar to the cultivated Washington palm but has no prickles on the petioles. Flowering March–April; flowers white or greenish; fruit a one-seeded, black berry.

Found on flatlands along the Rio Grande in the extreme southern Rio Grande Plains (Cameron County) and used as an ornamental northward to the southern Coastal Prairie.

PHYTOLACCACEAE

SNAKE-EYES, PUTIA, OJO DE VIBORA
Phaulothamnus spinescens Gray

Erect shrub to 3 m tall with dense, spiny branches. Leaves simple, alternate or clustered at the nodes, spatulate to oblanceolate, with entire margins. Flowering August–September; male and female flowers on separate plants; fruit a one-seeded, white, transparent drupe.

Found on clayey soils in the southern Rio Grande Plains. This plant has acquired the name snake-eyes because the transparent, fleshy fruit with the solitary, black seed seen within gives the appearance of a small eye.

RANUNCULACEAE

TEXAS VIRGIN'S-BOWER, BARBAS DE CHIVATO, OLD MAN'S BEARD
Clematis drummondii T. & G.

Perennial, clambering or straggling, delicately branched vine that is often shrublike. Leaves opposite, pinnately compound, and five to seven foliate. Flowering April–September; flowers white, male and female flowers on separate plants or polygamo-dioecious; fruit an achene that is borne in a conspicuous, globose head.

In dry soils, dry washes, and rocky canyons, commonly climbing over trees, shrubs, and along the ground in the Coastal Prairie, Rio Grande Plains, and Trans-Pecos.

The leaves are browsed by white-tailed deer.

RHAMNACEAE

HOGPLUM, TEXAS COLUBRINA, GUAYULE

Colubrina texensis (T. & G.) Gray

Rounded, thicket-forming shrub 1–2 m tall with light gray bark and twigs. Leaves simple, alternate or clustered at the nodes, ovate with toothed margins. Flowering late spring–early summer; flowers greenish yellow; fruit a brown or black drupe separating into two or three nutlets.

Found on gravelly or rocky slopes and along washes and arroyos in the Trans-Pecos, Edwards Plateau, Rio Grande Plains, and Coastal Prairie.

White-tailed deer browse the leaves. Javelina and several species of birds consume the fruit. It is poisonous to sheep.

BLUEWOOD CONDALIA, BRASIL, CAPUL NEGRO

Condalia hookeri M. C. Johnst.

Thicket-forming, spiny shrub or tree to 9 m tall. Leaves alternate or fascicled on short spiny branches, obovate with entire margins. Flowering in summer; flowers very small, greenish; fruit a black drupe when mature.

Found in the Trans-Pecos, southern Edwards Plateau, Rio Grande Plains, and Coastal Prairie.

The leaves, which contain about 15 percent crude protein, are frequently browsed by white-tailed deer. The fruits are eaten by the coyote, raccoon, bobwhite and scaled quail, and several species of passerine birds. It also provides cover for wildlife. The fruits are edible, sweet, and succulent.

KNIFE-LEAF CONDALIA, SQUAW-BUSH, COSTILLA

Condalia spathulata Gray

Rounded shrub usually less than 1 m tall with very spiny, grayish green branches and small leaves. Leaves simple, alternate or in fascicles at short shoots, spatulate with entire margins. Flowering summer–fall; flowers very small, greenish; fruit a small, dark red or black drupe.

Found on dry, rocky hillsides and along stony arroyos in the western Rio Grande Plains and southern Edwards Plateau.

The young shoots are occasionally browsed by livestock and white-tailed deer.

COYOTILLO

Karwinskia humboldtiana (R. & S.) Zucc.

Unarmed shrub or small tree usually 1–2 m tall. Leaves dark green above and paler beneath, opposite, oblong or elliptic-oblong, with entire margins. Flowering summer–fall; flowers small, greenish; fruit a brown or black drupe.

Found in the Trans-Pecos, southern Edwards Plateau, and Rio Grande Plains.

The seeds and leaves are poisonous to cattle, sheep, goats, horses, and swine. The fruits are eaten by the coyote and chachalaca.

LOTEBUSH, GUMDROP TREE, CLEPE
Ziziphus obtusifolia (T. & G.) Gray

Spiny, much-branched shrub 1–2 m tall with grayish green branches and twigs. Leaves alternate, very variable in shape from deltoid to ovate or oblong to nearly linear, with entire or very coarsely toothed margins. Flowering in summer; flowers very small, greenish; fruit a black drupe.

Locally abundant throughout the state except for east Texas Pineywoods and the higher parts of the High and Rolling Plains.

The leaves are occasionally browsed by white-tailed deer, and the fruits are eaten by the gray fox, raccoon, coyote, and chachalaca.

ROSACEAE

PEACH BUSH, DURAZNO, DURASNILLO
Prunus texana Dietr.

A dwarf shrub to 1 m tall with crooked branches and subevergreen leaves. The young branchlets are light gray and pubescent. Leaves simple, alternate or fascicled, oblong to ovate-oblong or elliptic, with toothed margins. Both upper and lower leaf surfaces are densely pubescent. Flowering in late winter–early spring; flowers very small, pink or white; fruit a one-seeded, velvety drupe.

An infrequent shrub usually found in sandy or sandy loam soils of the Rio Grande Plains and Edwards Plateau.

The fruits are edible, peachlike, and make excellent preserves.

RUBIACEAE

COMMON BUTTONBUSH, HONEY-BALLS, GLOBE-FLOWERS
Cephalanthus occidentalis L.

Shrub or small tree to 15 m tall, often swollen at the base. Leaves large, simple, opposite or ternate, ovate to ovate-oblong or narrowly laceolate, with entire margins. Flowering June–September; flowers white, very showy; fruit a round cluster of reddish brown nutlets.

Found in swamps, margins of ponds, streams, and other low areas throughout Texas.

It is frequently cultivated as an ornamental and provides a good food for bees. The dry nutlets are eaten by a number of water birds, but are reportedly toxic to other animals. It is frequently used as a nesting site by colonial water birds.

MEXICAN BUTTONBUSH
Cephalanthus salicifolius H. & B.

Shrub or small tree attaining a height of 3–4 m. Leaves simple, opposite and ternate, narrowly oblong to elliptic-oblong or laceolate, with entire margins. Flowering March–July; flowers white, very showy; fruit a round cluster of brown nutlets.

Rare in wet soil near the Rio Grande in the extreme southern Rio Grande Plains (Hidalgo County).

CRUCILLO, TEXAS RANDIA
Randia rhagocarpa Standl.

Rigid, thorny shrub attaining a height of 3–4 m. The crosslike, paired thorns are a conspicuous characteristic of this plant. Leaves clustered at the nodes, opposite, broadly obovate or oval, with entire margins. Flowering February–June; flowers pale green; fruit a one-seeded, black drupe.

Infreqent in open chaparral in the southern Rio Grande Plains (Cameron, Hidalgo, Starr, and Willacy counties).

RUTACEAE

MEXICAN AMYRIS, MOUNTAIN TORCHWOOD AMYRIS
Amyris madrensis Wats.

Slender shrub or small tree usually 1–3 m tall with all parts citrus-scented when bruised. Leaves opposite or subopposite, pinnately compound with five to nine pairs of leaflets (leaflets ovate or rhombic-ovate, with entire or toothed margins). Flowering spring–fall; flowers green or white; fruit a one-seeded drupe.

A rare species found in chaparral in the extreme southern Rio Grande Plains (Cameron, Hidalgo, and Willacy counties).

CHAPOTILLO, TEXAS TORCHWOOD AMYRIS

Amyris texana (Buckl.) P. Wilson

Much-branched, aromatic, rounded shrub often only 1 m tall, occasionally to 2 m. Leaves alternate, trifoliolate (leaflets elliptic to ovate) with toothed margins. Flowering spring–fall; flowers greenish white; fruit a one-seeded drupe.

Locally abundant in chaparral in the extreme southern Rio Grande Plains and north along the Coastal Prairies and Marshes to Matagorda County.

This plant is generally a low undershrub and is usually mixed with other dense chaparral. White-tailed deer occasionally browse the leaves.

JOPOY

Esenbeckia berlandieri Baill.

Small tree 3–6 m tall with whitish bark and a rounded top. Leaves alternate, trifoliolate (leaflets elliptic with entire margins), glossy and dark green. Flowering summer–fall; flowers creamy white; fruit a five-lobed (rarely three- or four-lobed) capsule.

Exceedingly rare, known from three trees in Cameron County in the extreme southern Rio Grande Plains. It can be found in cultivation at several residences in Cameron and Hidalgo counties.

BARETTA

Helietta parvifolia (Gray) Benth.

Slender, evergreen shrub or small tree 2–4 m tall. Branches erect and spineless forming an irregularly shaped small crown. Like other members of the citrus family, this species is aromatic. Leaves opposite and trifoliolate (leaflets usually oblong-obovate with entire margins). Flowering in spring; flowers greenish white; fruit three to four samaralike carpels, separating at maturity.

Found in thickets on gravel and rocky hills a few miles east of Rio Grande City (Starr County) in the extreme southern Rio Grande Plains where originally it was more abundant. This species is threatened with extinction in the United States because of the clearing of chaparral.

PEPPERBARK, HERCULES-CLUB, PRICKLY ASH, TICKLE-TONGUE, TOOTHACHE TREE

Zanthoxylum clava-herculis L.

Aromatic shrub or small tree with a broad, rounded crown and attaining a height of 3–4 m. Leaves alternate, odd-pinnately compound with nine to 17 leaflets having toothed margins (leaflets ovate or ovate-lanceolate). Flowering in early spring; flowers yellow green; fruit a brownish follicle (two to five together).

A frequent shrub in the east Texas Pinewoods and southeast Texas Coastal Prairie (San Patricio County), infrequent in the Edwards Plateau and northeastern Rio Grande Plains.

LIME PRICKLYASH, COLIMA
Zanthoxylum fagara (L.) Sarg.

Aromatic, very prickly, evergreen shrub or small tree to 9 m tall. Leaves alternate, odd-pinnately compound with five to 13 leaflets on a broadly winged rachis (leaflets obovate or oval, with toothed margins). Flowering winter–spring; flowers small, yellowish green; fruit a one-seeded, rusty brown follicle.

Frequent in chaparral in the Rio Grande Plains and Coastal Prairie.

The leaves (which contain about 15 percent crude protein) and soft twigs are an important food source for white-tailed deer. It is often used as a nesting site for several species of passerine birds.

TICKLE-TONGUE, TOOTHACHE TREE, PRICKLY ASH
Zanthoxylum hirsutum Buckl.

Thorny shrub or small tree 1–4 m tall, very aromatic. Leaves alternate, odd-pinnately compound with three to seven leaflets having crinkled margins (leaflets elliptic to oblong or oval). Flowering in early spring; flowers greenish; fruit a reddish brown capsule.

Frequent in sandy soils of the Rio Grande Plains, brushy areas of the Edwards Plateau, lower parts of the High and Rolling Plains, and Edwards Plateau.

White-tailed deer occasionally browse the leaves.

SALICACEAE

EASTERN COTTONWOOD, ALAMO
Populus deltoides Marsh.

Tree to 30 m tall or more with a rather short, thick trunk, heavy branches, and a wide, open crown. Leaves large, simple, alternate, deciduous, deltoid to deltoid-ovate or suborbicular-ovate, with toothed margins. Flowering March–July; male and female flowers separate on the same plant; fruit a capsule.

Found along practically every watercourse of any size and about most springs and waterholes in the Pineywoods, Edwards Plateau, and northern Rio Grande Plains.

The seeds are eaten by birds and the foliage is browsed by cattle. This tree is well known for the cottony seeds it sheds when the fruit is ripe. It is often planted as an ornamental.

TEXAS SANDBAR WILLOW, SILVER-FRUITED SANDBAR WILLOW
Salix exigua Nutt.
[syn. *S. interior* var. *angustissima* (Anderss.) Dayton]

A slender, upright shrub or small tree attaining a height of 8 m. Leaves deciduous, alternate, linear-lanceolate with toothed margins. Flowering April–May; male and female flowers on separate plants; fruit a brownish capsule.

Found along streams and near bodies of water in the Rio Grande Plains and Trans-Pecos.

SAUZ, BLACK WILLOW
Salix nigra Marsh.

A fast-growing tree attaining a height of 20 m or more, sometimes with several trunks. Leaves deciduous, simple, alternate, narrowly lanceolate with toothed margins. Flowering April–May; flowers cream-colored, male and female flowers on separate plants; fruit a small, light brown capsule.

Frequent in alluvial soils along streams and about bodies of water throughout the eastern two-thirds of Texas (throughout the Rio Grande Plains and southern Coastal Prairie, including the barrier islands).

This is the willow that sheds the white, cottonlike substance with the ripening of the fruit. White-tailed deer occasionally browse the leaves. Several species of birds use this tree for nesting sites.

SAPINDACEAE

JABONCILLO, WESTERN SOAPBERRY
Sapindus drummondii Hook. and Arn.
[syn. *S. saponaria* L. var. *drummondii* (Hook. and Arn.) L. Benson]

Trees attaining a height of 15 m with erect branches usually forming a rounded crown. Leaves deciduous, alternate, abruptly pinnate, with four to 11 pairs of leaflets (leaflets elliptic-lanceolate to narrowly lanceolate-attenuate, with entire margins). Flowering March–July; flowers white; fruit globular, fleshy, from white to yellowish to blackish with one seed.

Scattered throughout Texas on moist soils on the edge of woods, along streams, and fencerows.

This tree is often cultivated.

SAPOTACEAE

LA COMA, COMA

Bumelia celastrina H.B.K.

A spiny shrub or small tree 2–9 m tall. Leaves dark green, alternate or fascicled at the nodes, oblanceolate to obovate or sometimes nearly elliptic, with entire margins. Flowering May–November; flowers greenish white; fruit a one-seeded, blue-black drupe.

Found along resacas, on gravelly hills, thickets, and salt marshes of the Rio Grande Plains.

The leaves contain 12 to 14 percent crude protein and are important food for white-tailed deer. The fleshy fruits are eaten by the chachalaca, raccoon, and coyote. It provides cover for wildlife and a nesting site for several species of passerine birds. The fruits are sweet and quite tasty.

WOOLLYBUCKET BUMELIA, COMA, CHITTIMWOOD

Bumelia lanuginosa (Michx.) Pers. var. *rigida* Gray

Shrub or irregularly shaped tree attaining a height of 15 m with stiff, more or less spiny branchelets. Leaves alternate or clustered, oblanceolate to obovate or elliptic, with entire margins. The leaves are covered with white pubescence on the lower surface. Flowering April–June; flowers white; fruit a purplish black drupe.

Found on limestone or gypseous soils, especially along bluffs of rivers and streams in the Edwards Plateau, northern Rio Grande Plains, and southern Coastal Prairie.

The fruit is eaten by birds; white-tailed deer browse the leaves and fruits.

SCROPHULARIACEAE

CENIZO, PURPLE SAGE, TEXAS SILVERLEAF
Leucophyllum frutescens (Berl.) I. M. Johnst.

Unarmed shrub to about 25 dm high, conspicuous because of the ash gray leaves. Leaves simple, alternate or clustered, elliptic-obovate with entire margins. Flowering throughout the year; flowers pale violet to purple or pink to white; fruit a two-valved, many-seeded capsule.

Frequent on rocky limestone hills, caliche cuestas, bluffs, ravines, arroyos, and chaparral in the Rio Grande Plains and southern Trans-Pecos.

This shrub is cultivated as an ornamental and is planted along Texas highways. The leaves are occasionally browsed by white-tailed deer.

SIMAROUBACEAE

ALLTHORN GOAT-BUSH, AMARGOSA
Castela texana (T. & G.) Rose

Low, much-branched, spiny shrub attaining a height of 2 m. Leaves simple, alternate or fascicled at the nodes, linear-oblong to lanceolate or narrowly oblanceolate, with entire margins. The leaves are shiny green above and silver-pubescent beneath. Flowering March–May; flowers very small, red or orange; fruit a one-seeded, bright red drupe.

Found on gravelly hills, chaparral thickets, gulf shores, and mesquite prairies in the southern Coastal Prairie, Rio Grande Plains, southern Edwards Plateau, and Trans-Pecos.

White-tailed deer browse the leaves and fruit. It is frequently used as a nesting site by several species of birds.

SOLANACEAE

BUSH PEPPER, BIRD PEPPER, CHILIPIQUIN
Capsicum annuum L. var. *glabriusculum* (Dunal) Heiser & Pickersgill

A low undershrub attaining a height of 3 m with slender, brittle, green branches and a broad, spreading top. Leaves simple, ovate to elliptic-lanceolate or lanceolate, with entire or slightly toothed margins. The upper leaf surfaces are dull green, whereas the lower surfaces are paler. Flowering throughout the year; flowers white; fruit a many-seeded, orange or red berry.

Found on ledges along rivers and in thickets and groves along arroyos in the Edwards Plateau, Rio Grande Plains, and southern Coastal Prairie.

A seasoning is made by soaking the berries in vinegar, and the berries make a spicy addition to local recipes. Several species of birds, including the Rio Grande turkey, eat the berries.

BERLANDIER WOLFBERRY, CILINDRILLO
Lycium berlandieri Dun.

Spiny, sparingly branched shrub attaining a height of 25 dm. Leaves alternate or fascicled at the nodes, linear to elliptic-spatulate, with entire margins. Flowering February–October; flowers blue to lavender or white; fruit a many-seeded, red berry.

Found on gravelly-rocky hills, limestone and clay flats, alkali flats, arroyos, scrubland and thickets in the Trans-Pecos, Rio Grande Plains, and Coastal Prairie.

The leaves are browsed by white-tailed deer, and several species of birds and mammals eat the fruit including the chachalaca and raccoon.

CAROLINA WOLFBERRY

Lycium carolinianum Walt. var. *quadrifidium* (Dun.) C. L. Hitchc.

Spiny, sparingly branched, erect or semitrailing shrub attaining a height of 1 m. Leaves thick, fleshy, alternate or fascicled at the nodes, linear to terete-spatulate, with entire margins. Flowering January–November; flowers lavender to purple; fruit a many-seeded, red berry.

Found about ponds, ditches, marshes, on wet, clay flats and salt flats, and in sandy gravelly soil on chaparral-covered hills in the Coastal Prairies and Marshes and Rio Grande Plains.

The fruit is eaten by several species of birds, including whooping cranes, and white-tailed deer occasionally browse the leaves.

TREE TOBACCO, MUSTARD TREE, RAPE, BUENA MOZA, GIGANTE

Nicotiana glauca Grah.

Evergreen shrub or small tree to 5 m tall. Leaves large, leathery, simple, alternate, ovate to oblong-lanceolate, with entire or undulated margins. Flowering throughout the year; flowers green to yellowish; fruit a two-celled, many-seeded capsule.

Found along streams, ditches, roadsides, and waste places in the southern Rio Grande Plains, southern Coastal Prairie, and Trans-Pecos.

This plant is a native of South America that has become naturalized in Texas. Tree tobacco is poisonous to cattle, sheep, and horses.

POTATO TREE, SALVADORA

Solanum erianthum D. Don.

Shrub or small tree to 3 m tall but usually only an irregular, open, flat-topped shrub. Leaves large, alternate, ovate to elliptic or oblong, with entire or undulated margins. Flowering April–October; flowers white; fruit a many-seeded, yellow berry.

Rare in open woods and thickets in the extreme southern Rio Grande Plains (as far north as Kenedy County).

TEXAS NIGHTSHADE, HIERBA MORA

Solanum triquetrum Cav.

Small, shrublike perennial attaining a maximum height of 2 m but usually much smaller. This plant is mostly herbaceous above but is somewhat woody toward the base. Leaves dark green, deciduous, simple, alternate, deltoid-cordate with entire margins. Flowering throughout the year; flowers white or tinged violet; fruit a red berry.

Found on low hills, slopes, in thickets and on breaks, and along fencerows in the Rio Grande Plains, southern Coastal Prairie, and Trans-Pecos.

The foliage is eaten by white-tailed deer and javelina.

STERCULIACEAE

AYENIA
Ayenia limitaris Cristóbal

Small shrub attaining a height of 15 dm with the young branches covered with soft white pubescence. Leaves in clusters at nodes, ovate with coarsely toothed margins. Flowering throughout the year; flowers small, yellow green; fruit a hairy, five-lobed capsule.

Rare in clay soil on the edge of thickets and in chaparral in the extreme southern Rio Grande Plains (Cameron and Hidalgo counties).

BROOM-WOOD, WOOLLY PYRAMID-BUSH
Melochia tomentosa L.

Shrub to 2 m tall, herbaceous above and woody below. The stems and branches are covered with fine pubescence. Leaves simple, alternate, rhombic-ovate to oblong or linear-lanceolate, with toothed margins. Flowering summer–fall; flowers pink or violet; fruit a five-celled, five-seeded capsule.

An infrequent species in open woodlands and chaparral in the southern Rio Grande Plains.

HIERBA DEL SOLDADO

Waltheria indica L.

Small subshrub or strong perennial herb with a woody base, attaining a maximum height of 2 m, although usually much shorter. Leaves alternate, ovate to orbicular or oblong, with toothed margins. The leaves are covered with pubescence. Flowering winter and spring; flowers yellow; fruit a small, one-seeded capsule.

Found in rocky or sandy soil in the extreme southern Rio Grande Plains (Hidalgo and Starr counties).

White-tailed deer browse the leaves.

TAMARICACEAE

SALTCEDAR, ATHEL

Tamarix aphylla (L.) Karst.

Reddish brown- to gray-barked tree attaining a height of 15 m. Leaves alternate, linear, vaginate, entire, and scalelike. Flowering in summer; flowers white; fruit a small, many-seeded capsule.

A native of Africa and the Middle East that has been introduced into Texas. This plant has become naturalized along streams, salt flats, and in waste places throughout the state. It is planted for windbreaks and sand binding and is also used as an ornamental.

SALTCEDAR, TAMARISK

Tamarisk chinesis Lour.

Shrub or tree to 8 m tall with brown to black purple bark. Leaves alternate, sessile, deltoid to lanceolate, entire, and scalelike. Flowering in summer; flowers white pink; fruit a small, many-seeded capsule.

Native of the Far East, introduced to the United States and naturalized throughout the southwest. Found in salt flats and waste places in the western half of Texas.

TURNERACEAE

DAMIANA, HIERBA DEL VENADO

Turnera diffusa Willd. var. *aphrodisiaca* (Ward) Urban

Small, aromatic shrub attaining a height of 2 m, although usually much smaller. Leaves simple, alternate or clustered, oblong-elliptic to elliptic-oblanceolate, with coarsely toothed margins. The upper surface of the leaves is smooth and light olive green and the lower surface is pubescent and whitish. Flowering throughout the year; flowers yellow; fruit a small, three-valved capsule with curved seeds.

Infrequent or rare on dry, chaparral-covered hillsides in the western Rio Grande Plains (usually near the Rio Grande).

ULMACEAE

PALO BLANCO, TEXAS SUGARBERRY
Celtis laevigata Willd.

Tree to 30 m tall with many stout, spreading branches forming a broad crown. Numerous corky warts are found on the bark of the trunk and the main branches. Leaves simple, alternate, lanceolate to oblong-lanceolate or sometimes ovate-lanceolate, with entire or partially toothed margins. Flowering March–April; flowers small, greenish, monoecious-polygamous; fruit a one-seeded, orange-red to black drupe.

Found in the eastern two-thirds of Texas. Frequent along streams, in palm groves, and in thickets in the Rio Grande Plains and southern Coastal Prairie.

The leaves are occasionally browsed by white-tailed deer and other mammals, and several species of birds eat the fruit including the cardinal, mockingbird, and chachalaca.

GRANJENO, DESERT HACKBERRY, SPINY HACKBERRY
Celtis pallida Torr.

Spiny, spreading evergreen shrub rarely over 3 m tall. Leaves simple, alternate, ovate to ovate-oblong or elliptic, with entire or coarsely toothed margins. Flowering in spring; flowers polygamous or monoecious, greenish white; fruit a one-seeded, yellow or orange drupe.

Found in the Trans-Pecos, Edwards Plateau, Rio Grande Plains, and Coastal Prairie.

Granjeno is a good food for wildlife; the fruits are eaten by a number of different birds and mammals including the cactus wren, cardinal, pyrrhuloxia, mockingbird, scaled quail, raccoon, coyote, and jackrabbit. The leaves and stems are heavily browsed by white-tailed deer. Both the fruit and leaves contain about 20 percent crude protein. The fruits are edible, sweet, and succulent.

NETLEAF HACKBERRY, PALO BLANCO

Celtis reticulata Torr.

Shrubs or small trees attaining a height of 16 m. Numerous corky warts are found on the bark of the trunk and main branches. Leaves simple, alternate, ovate with entire or partially toothed margins. This species can be distinguished from *Celtis laevigata* by its leaves, which are conspicuously reticulate-veined beneath and more hispid to the touch. Flowering in spring; flowers small, greenish, monoecious-polygamous; fruit a one-seeded, reddish or reddish black drupe.

Found on limestone hills, breaks and rocky canyon slopes, and in arroyos, mesquite groves, and about tanks, ponds, and along watercourses in the northern Rio Grande Plains, Coastal Prairies, Edwards Plateau, and Trans-Pecos.

The fruits are eaten by several species of birds including the cardinal, pyrrhuloxia, and mockingbird and mammals including the raccoon.

CEDAR ELM, OLMO

Ulmus crassifolia Nutt.

Tree attaining a height of 25 m with twigs or branches having two opposite corky wings. Leaves simple, alternate, ovate to ovate-oblong or elliptic, with toothed margins. Flowering July–October; flowers red to green; fruit a small, green, pubescent samara.

In woodlands, ravines, and open slopes in the Edwards Plateau, Rio Grande Plains, and southern Coastal Prairie.

This tree provides cover and nesting sites for several species of passerine birds.

VERBENACEAE

WHITEBRUSH, COMMON BEE-BRUSH, JAZMINILLO

Aloysia gratissima (Gill. & Hook.) Troncoso.

Aromatic, slender, thicket-forming shrub attaining a height of 3 m with many stiff, gray branches. Leaves opposite, narrow-oblong or elliptic to lanceolate-oblong, with entire or toothed margins. Flowering March–November, especially after rains; flowers white; fruit a small drupe containing two nutlets.

Found in sandy soil, gravelly hillsides, chaparral, thickets, arroyos, and limestone bluffs of the Trans-Pecos, Edwards Plateau, and Rio Grande Plains.

Whitebrush is poisonous to horses, mules, and burros. It is an important nectar-producing plant for honey and provides cover for wildlife.

SWEET-STEM, VARA DULCE

Aloysia macrostachya (Torr.) Moldenke

Aromatic, erect shrub attaining a height of 2 m with numerous branches. Leaves simple, opposite, ovate, pubescent with partially toothed margins. Flowering January–October; flowers pink to red or lavender; fruit a small drupe separating into two nutlets.

On rocky hillsides and in dry gravelly arroyos in the southern Rio Grande Plains.

BEAUTYBERRY, FILIGRANA
Callicarpa americana L.

Much-branched shrub to 3 m tall with branches and leaves covered with dense, matted pubescence. Leaves large, simple, opposite or ternate, ovate to elliptic, with coarsely-toothed margins except near the ends. Flowering June–December; flowers bluish, pinkish, reddish or white; fruit a rose to purple or violet to blue, berrylike drupe.

Found in Coastal Prairies and Marshes of southeast Texas (southwest to San Patricio and Nueces counties) and inland to parts of the Edwards Plateau.

The fruit is eaten by several species of birds and mammals including the bobwhite quail, raccoon, opossum, and gray fox.

NEGRITO, ORCAJUELA, ENCORBA GALLINA, BERLANDIER FIDDLEWOOD
Citharexylum berlandieri Robins.

Shrub or small tree to 6 m tall with heavy, gray branches. Leaves opposite, lanceolate or oblong to elliptic or ovate, with entire or partially toothed margins. Flowering February–August; flowers white; fruit a two-seeded, yellow or red drupe (blackening with drying).

Infrequent to rare in thickets, on flats, hillsides, and semidesert roadsides in the southern Rio Grande Plains (Cameron, Hidalgo, and Willacy counties).

BOXTHORN FIDDLEWOOD, MISSION FIDDLEWOOD

Citharexylum brachyanthum (Gray) Gray

Slender shrub rarely more than 18 dm tall with long, slender, flexible branches. Leaves small, opposite on young shoots and clustered on short, spurlike branchlets on older wood, spatulate with entire margins. Flowering April–June, later following rains; flowers white; fruit a two-seeded, orange red drupe.

Found in chaparral in the extreme southern Rio Grande Plains (Hidalgo and Starr counties).

The fruits are eaten by several species of birds including the scaled quail and cardinal.

WEST INDIAN LANTANA, AFROMBRILLA, HEDIONA

Lantana camara L.

Branching shrub attaining a height of 2 m with stems and branches usually unarmed or with only comparatively few weak prickles. Leaves opposite, ovate to oblong, with toothed margins. Flowering spring–fall; flowers red to yellow, white, lilac, or rose; fruit a black drupe containing two nutlets.

This plant is widely cultivated but escapes and grows wild in the Edwards Plateau and Rio Grande Plains. It is found in the West Indies, Central and South America, Asia, and Africa.

The leaves are poisonous to livestock.

TEXAS LANTANA, HIERBA DE CRISTO, CALICO BUSH

Lantana horrida H.B.K.

Aromatic, much-branched shrub attaining a height of 2 m; branches unarmed or with stout recurved prickles. Leaves simple, opposite, ovate to broadly ovate, with coarsely toothed margins. The upper leaf surface is hairy. Flowering spring–fall; flowers yellow to orange or red; fruit a black or dark blue drupe containing two nutlets.

Found almost throughout Texas except the northwest part. Usually found on sandy soils in the Rio Grande Plains.

Bobwhite quail occasionally eat the fruit.

DESERT LANTANA, HIERBA NEGRA, MEJORANA, YERBA DE CRISTO

Lantana macropoda Torr.

Pubescent, aromatic shrub with slender stems and branches, usually 3–9 dm tall. Leaves simple, opposite, ovate or lanceolate, with sharply toothed margins. Flowering February–May; flowers white or pink, often with yellow centers; fruit a thin-fleshed drupe, dividing into two nutlets.

Found on dry soil of gravelly hills and rocky slopes, along arroyos, roadsides, and waste places in the Edwards Plateau, Trans-Pecos, southern Coastal Prairie, and Rio Grande Plains.

White-tailed deer occasionally browse this small shrub.

BUSHY LIPPIA, HIERBA NEGRA, HIERBA DEL NEGRO

Lippia alba (Mill.) N. E. Brown

Much-branched, aromatic shrub to 2 m tall with long, rooting suckers at the base. Leaves opposite or ternate, ovate or oblong, margins toothed but entire toward the base, upper and lower surfaces pubescent. Flowering March–October; flowers purple to violet, pink or white; fruit a dry drupe containing two nutlets.

Found in woods, river banks, and resacas in the southern Rio Grande Plains (Cameron and Hidalgo counties) and northeast to Wharton County along the Coastal Prairie.

This species is widely cultivated and has been introduced in many areas.

REDBRUSH LIPPIA, HIERBA DULCE, OREGANO CIMARRON

Lippia graveolens H.B.K.

Slender, aromatic, pubescent shrub attaining a height of 3 m. Leaves simple, opposite, oblong or ovate-oblong to elliptic, with small blunt teeth on the margins. Flowering March–December; flowers yellowish white with yellow centers; fruit a small, dry drupe dividing into two nutlets.

Found on dry, rocky hills, valleys, arroyos, chaparral, and open desert scrub in the Trans-Pecos, Rio Grande Plains, and northeast to Austin and Houston counties.

ZYGOPHYLLACEAE

CREOSOTE BUSH, GOBERNADORA, HEDIONDILLA
Larrea tridentata (DC.) Coville

Evergreen, aromatic shrub to 3 m tall with an odor of creosote (especially when moist). Leaves resinous, opposite, bifoliate with two leaflets (ovate to oblong or obovate). Flowering February–August; flowers yellow; fruit a whitish-pubescent, five-celled capsule.

Occasional on shallow ridges (calcareous soils) in Webb, Jim Hogg, and Zapata counties of the Rio Grande Plains, becoming very prevalent in the Trans-Pecos and western Edwards Plateau.

This plant is poisonous to sheep.

GUAYACÁN, SOAP-BUSH, IRON-WOOD, TEXAS PORLIERIA
Guaiacum angustifolium Engelm.

[syn. *Porlieria angustifolia* (Engelm.) Gray]

Evergreen shrub or small tree to 7 m tall, often growing in clumps; the branches black, thick, and stubby. Leaves opposite or crowded in fascicles at the nodes, pinnately compound with four to eight pairs of leaflets (leaflets linear-oblong to linear-spatulate, with entire margins). Flowering March–April; flowers violet or purple; fruit a brown capsule containing one to three beanlike red, yellow, or orange seeds.

Frequent throughout the Rio Grande Plains, southern Edwards Plateau, and Trans-Pecos; rare on well-drained sites in the Coastal Prairie.

Guayacan is an important browse plant for deer. The leaves contain 16 to 18 percent crude protein.

BIBLIOGRAPHY

Alaniz, M. A., and J. H. Everitt. 1978. Germination of Texas ebony seed. J. Rio Grande Valley Hort. Soc., 32:95-100.

——. 1980. Germination of Anacua seeds. J. Rio Grande Valley Hort. Soc., 34:75-80.

Arnold, L. A., Jr., and D. L. Drawe. 1979. Seasonal food habits of white-tailed deer in the South Texas Plains. J. Range Manage., 32:175-178.

Benson, L. 1959. Plant classification. D. C. Heath and Co., Boston. 901 pp.

Chamrad, A. D. 1966. Winter and spring food habits of white-tailed deer on the Welder Wildlife Refuge. Unpublished M.S. thesis, Texas Technological Coll., Lubbock. 181 pp.

Chamrad, A. D., and T. W. Box. 1968. Food habits of white-tailed deer in South Texas. J. Range Manage., 21:158-164.

Clover, E. U. 1937. Vegetational survey of the lower Rio Grande Valley, Texas. Madroño, 4:41-66; 77-100.

Correll, D. S., and M. C. Johnston. 1970. Manual of the vascular plants of Texas. Univ. of Texas, Dallas. 1881 pp.

Dallas Morning News, The. 1986. The Texas Almanac. Dallas. 716 pp.

Davis, R. B. 1951. The food habits of white-tailed deer on the cattle stocked, live oak-mesquite ranges of the King Ranch, as determined by analyses of deer rumen contents. Unpublished M.S. thesis, Texas A&M Coll., College Station. 97 pp.

——. 1952. A study of some interrelationships of a native south Texas range, its cattle, and its deer. Unpublished Ph.D. dissertation, Texas A&M Coll., College Station. 114 pp.

Davis, R. B., and C. K. Winkler. 1968. Brush vs cleared range as deer habitat in southern Texas. J. Wildl. Manage., 32:321-329.

Davis, W. B. 1974. The mammals of Texas. Texas Parks and Wildlife Department, Bull. 41, rev. ed., Austin. 267 pp.

Drawe, D. L.. 1967. Seasonal forage preferences of deer and cattle on the Welder Wildlife Refuge. Unpublished M.S. thesis, Texas Technological Coll., Lubbock. 97 pp.

——. 1968. Mid-summer diet of deer on the Welder Wildlife Refuge. J. Range Manage., 21:164-166.

Drawe, D. L., and T. W. Box. 1968. Forage ratings for deer and cattle on the Welder Wildlife Refuge. J. Range Manage., 21:225-228.

Everitt, J. H. 1972. Spring food habits of white-tailed deer (*Odocoileus virginianus* Boddaert) on the Zachry Ranch in south Texas. Unpublished M.S. thesis, Texas A&I Univ., Kingsville. 114 pp.

——. 1986. Nutritive value of fruits or seeds of 14 shrub and herb species from south Texas. Southwestern Nat., 31:101-104.

Everitt, J. H., and M. A. Alaniz. 1980. Fall and winter diets of feral pigs in south Texas. J. Range Manage., 33:126-129.

——. 1981. Nutrient content of cactus and woody plant fruits eaten by birds and mammals in south Texas. Southwestern Nat., 26:301-305.

Everitt, J. H. and D. L. Drawe. 1974. Spring food habits of white-tailed deer in the South Texas Plains. J. Range Manage., 27:15-20.

Everitt, J. H., and C. L. Gonzalez. 1976. Discovery of candelilla (*Euphorbia antisyphilitica* Zucc.) in the lower Rio Grande Valley of Texas. J. Rio Grande Valley Hort. Soc., 30:109-111.

——. 1979. Botanical composition and nutrient content of fall and early winter diets of white-tailed deer in south Texas. Southwestern Nat., 24:297-310.

——. 1981. Seasonal nutrient content in food plants of white-tailed deer on the South Texas Plains. J. Range Manage., 34:506-510.

Everitt, J. H., C. L. Gonzalez, M. A. Alaniz, and G. V. Latigo. 1981. Food habits of the collared peccary on south Texas rangelands. J. Range Manage., 34:141-144.

Everitt, J. H., C. L. Gonzalez, G. Scott, and B. E. Dahl. 1981. Seasonal food preferences of cattle on native range in the South Texas Plains. J. Range Manage., 34:384-388.

Everitt, J. H., F. S. Lentz, and L. Lentz. 1990. Additional records of *Capparis incana* (Capparidaceae) in extreme southern Texas. Southwest Nat., 35:370–371.

Gould, F. W. 1975. Texas plants—A checklist and ecological summary. MP-585, Texas Agricultural Experiment Station, Texas A&M Univ., College Station, 121 pp.

Guerrero, E. J. Undated. Scientific, standard, and spanish names of woody plants in south Texas. USDA, Soil Conservation Service, Rio Grande City, Texas. 1 p.

Heep, M. R., and R. I. Lonard. 1986. *Esenbeckia berlandieri* (Rutaceae) rediscovered in extreme southern Texas. Southwestern Nat., 31:259-260.

Higginbotham, I., Jr. 1975. Composition and production of vegetation on the Zachry Ranch in the South Texas Plains. Unpublished M.S. thesis, Texas A&I Univ., Kingsville. 131 pp.

Johnston, M. C. 1952. Vegetation of eastern Cameron County, Texas. Unpublished M.S. thesis, Univ. of Texas, Austin. 127 pp.

——. 1955. Vegetation of the eolian plain and associated coastal features of southern Texas. Unpublished Ph.D. dissertation, Univ. of Texas, Austin. 167 pp.

——. 1963. Past and present grasslands of southern Texas and northeastern Mexico. Ecology, 44:456-466.

Jones, F. B. 1975. Flora of the Texas coastal bend. Welder Wildlife Foundation Contribution B-6, Rob and Bessie Welder Wildlife Foundation, Sinton, Texas. 262 pp.

Jones, F. B., C. M. Rowell, Jr., and M. C. Johnston. 1961. Flowering plants and ferns of the Texas coastal bend counties. Welder Series B-1, Rob and Bessie Welder Wildlife Foundation, Sinton, Texas. 146 pp.

Jones, J. K., Jr., R. S. Hoffmann, D. W. Rice, C. Jones, R. J. Baker, and M. D. Engstrom. 1992. Revised checklist of North American mammals north of Mexico, 1991. Occas. Papers Mus., Texas Tech Univ., 146:1–23.

Lehmann, V. W. 1984. Bobwhites in the Rio Grande Plain of Texas. Texas A&M Univ. Press, College Station. 371 pp.

Lonard, R. I., J. H. Everitt, and F. W. Judd. 1991. Woody plants of the lower Rio Grande Valley, Texas. Texas Memorial Museum, Univ. of Texas Press, Austin. 179 pp.

Lonard, R. I. and F. W. Judd. 1980. Phytogeography of South Padre Island, Texas. Southwestern Nat, 25:313-322.

——. 1981. The terrestrial flora of South Padre Island, Texas. Texas Memorial Museum, Misc. Papers, 6:1-74.

Lonard, R. I., F. W. Judd, and S. L. Sides. 1978. Annotated checklist of the flowering plants of South Padre Island, Texas. Southwestern Nat, 23:497-510.

Marion, W. R. 1976. Plain Chachalaca food habits in south Texas. Auk 93:376-379.

Mayeux, H. S., Jr., and C. J. Scifres. 1978. Goldenweeds—new perennial range weed problems. Rangeman's J., 5:91-93.

Porter, C. L. 1959. Taxonomy of flowering plants. W. H. Freeman and Co., San Francisco. 452 pp.

Rappole, J. H., and G. W. Blacklock. 1985. Birds of the Texas Coastal Bend; abundance and distribution. Texas A&M Univ. Press, College Station. 126 pp.

Rare Plant Study Center. Spring 1974. Rare and endangered plants native to Texas. Univ. of Texas Press, Austin. 12 pp.

Runyon, R. 1947. Vernacular names of plants indigenous to the lower Rio Grande Valley of Texas. The Brownsville News Publishing Co., Brownsville, Texas. 24 pp.

Scifres, C. J. 1980. Brush management. Texas A&M Univ. Press, College Station. 360 pp.

Scifres, C. J., J. L. Mutz, and G. P. Durham. 1976. Range improvement following chaining of south Texas mixed brush. J. Range Manage, 29:418-421.

Sperry, O. E., J. W. Dollahite, G. O. Hoffman, and B. J. Camp. 1968. Texas plants poisonous to livestock. B-1028, Texas Agricultural Experiment Station, Texas A&M Univ., College Station, 57 pp.

Standley, P. C. 1920-1926. Trees and shrubs of Mexico. Contrib. U. S. Nat. Herb., 23:1-1721.

Vines, R. A. 1960. Trees, shrubs, and woody vines of the southwest. Univ. of Texas Press, Austin. 1104 pp.

Weniger, D. 1970. Cacti of the Southwest. Univ. of Texas Press, Austin. 249 pp.

Wills, M. M., and H. S. Irwin. 1961. Roadside flowers of Texas. Univ. of Texas Press, Austin. 295 pp.

GLOSSARY

Terms are according to Vines (1960) and Correll and Johnston (1970).

Abrupt Terminating suddenly.
Acaulescent Without a stem above ground.
Achene A small, dry, indehiscent, one-seeded, usually hard fruit in which the ovary wall is free from the seed.
Acorn The fruit of oaks.
Acuminate Gradually tapering into an extended point.
Aggregate Collected into dense clusters or tufts.
Alkali A soluble mineral salt present in some soils of arid regions.
Alkaline Having the properties of an alkali.
Alluvial Soils deposited by running water.
Alternate One leaf at a node.
Annual Of one year's duration from seed to maturity and death.
Aperture An opening, gap, cleft.
Apex The tip of an organ.
Appressed Closely and flatly pressed against.
Arching To form or bend into the shape.
Aromatic Fragrant, pungent, spicy to smell or taste.
Arroyo A water course, or channel or gully, often dry, carved by water.
Ascend To move upward.
Attenuate Slenderly tapering or prolonged.
Axil Upper angle formed by a leaf or branch with the stem.
Axillary Situated in the axil.
Axis The central stem along which parts or organs are arranged.

Basal Situated at or near the base.
Base The bottom of anything, considered as its support.
Berry A pulpy, indehiscent fruit usually containing numerous seeds.
Bifoliate A leaf composed of two leaflets.
Bipinnate Doubly or twice pinnate; when both primary and secondary divisions of a leaf are pinnate.
Bisexual Having both sexes on the same individual plant.
Bract A more or less modified leaf subtending a flower or belonging to a cluster of flowers.
Branch (branches) A stem growing from the trunk or from a bough of a tree or shrub; a limb.
Browse Tender shoots, twigs, and leaves, fit for food for livestock and wildlife.

Calcareous Containing an excess of available calcium.
Calyx The external, usually green, whorl of a flower, contrasted with the inner showy corolla.
Capsule A dry, dehiscent fruit composed of more than one carpel.
Carpel A simple pistil, or one of the modified leaves forming a compound pistil.
Chaparral A type of low scrub, commonly with dense, twiggy, thorny habit and evergreen leaves.

Cluster (clustered, clustering) A number of similar things growing together or of things collected together.
Coarse Harsh, rough, or rude as opposed to delicate or dainty.
Columnar Having the form of a column.
Compound Having two or more similar parts in one organ.
Compressed Flattened laterally.
Cone An inflorescence of flowers or fruit with overlapping scales.
Conical Cone-shaped with the point of attachment at the broad base.
Connate Said of similar parts that are united, at least at the base.
Conspicuous Obvious or prominent to the eye.
Cordate Heart-shaped with the notch at the base and ovate in general outline.
Corolla The inner perianth whorl of a flower.
Creeping Spreading over or beneath the ground and rooting at the nodes.
Creosote A heavy, oily, colorless liquid of strong odor obtained from wood tar.
Crimson A deep red color.
Crinkle To bend over without breaking off.
Crown Usually referring to the branches and foliage of a tree.
Cuestas A topographic ridge, usually a ridge that has a long, gentle slope on one side and a short, steep slope on the other side.
Cuneate Triangular with the narrow part at point of attachment.
Cylindrical Having a cylindric shape that is elongate and circular in cross section.

Deciduous Falling off, as petals fall after flowering or leaves of nonevergreen trees in autumn.
Dehiscent The bursting open of a capsule or pod at maturity.
Deltoid Equilaterally triangular.
Depressed Low, as if flattened from above.
Densely Having its parts massed or crowded together.
Di- Greek prefix meaning two or double.
Diffuse (diffusely) Scattered or spreading.
Dioecious Having staminate and pistillate flowers in different plants.
Drupe A fleshy or pulpy fruit in which the inner portion is hard and stony, enclosing the seed.

Elliptic In the form of a flattened circle usually more than twice as long as broad.
Entire Undivided; the margin continuous, not incised or toothed.
Epidermis The true cellular skin or covering of a plant below the cuticle.
Erect Upright in relation to the ground or sometimes perpendicular to the surface of attachment.
Evergreen Remaining green through the winter.

Fascicle (fascicled) A close bundle or cluster.
Filiform Threadlike.
Fleshy Thick and juicy.
Flexible Capable of being bent.
Floret A small flower of a dense cluster.
Foliate Referring to leaves as opposed to leaflets.
Foliage The leafy covering, especially of trees.
Follicle A dry, monocarpellary fruit, opening only on the ventral suture.

Gamopetalous Referring to the petals being more or less united.
Globose Rounded.
Glutinous Gluelike or sticky.
Gypseous Containing hydrous calcium sulfate.

Hairy Bearing or covered with hair.
Hastate (hastately) Of the shape of an arrow head but with the basal lobes turned outward.
Herb A plant without persistent woody stem, at least above ground.
Herbaceous Having the characteristics of an herb.
Hispid Rough with stiff or bristly hairs.

Inconspicuous Not prominent or striking.
Indehiscent Remaining closed at maturity.
Inflorescence The flower cluster of a plant; the disposition of flowers.
Infrequent Happening or occurring seldom.
Inundated To spread or flow over; to flood.
Irregular Showing a lack of uniformity.

Lanceolate Lance-shaped.
Leaflet A segment of a compound leaf.
Legume The seed vessel of Leguminosae, one-celled and two-valved, but various in form.
Linear Long and narrow.
Lobe (lobed) A division or segment or an organ.
Lyrate Lyre-shaped.

Margin A border or edge of the leaf.
Mono- Greek prefix meaning one or single.
Monoecious Having staminate and pistillate flowers on the same plant but not perfect ones.

Naturalize Of foreign origin but established and reproducing itself as though a native.
Node The joint of a stem; the point of insertion of a leaf or leaves.
Nutlet Any small and dry nutlike fruit or seed.

Oblanceolate Inversely lanceolate.
Oblique Of unequal sides; slanting.
Oblong Much longer than broad with nearly parallel sides.
Obovate Inversely ovate.
Opposite Opposed to each other, such as two opposite leaves at a node.
Orbicular Circular in outline.
Ornamental A plant cultivated essentially for decorative purposes.
Oval Broadly elliptic.
Ovary The female reproductive organ.
Ovate Egg-shaped with the broader end closer to the stem.

Palatable Agreeable to the taste.
Perianth The floral envelope, usually consisting of distinct calyx and corolla.

Perennial Lasting from year to year.

Petiole A leaf stalk.

Photosynthesis The process by which plant cells manufacture sugar from carbon dioxide and water in the presence of chlorophyll and sunlight.

Photosynthetic Refers to the process of photosynthesis.

Pinnae A leaflet or primary division of a pinnate leaf.

Pinnate A compound leaf having the leaflets arranged on each side of a common petiole; featherlike.

Pinnatifid Pinnately cleft into narrow lobes not reaching to midrib.

Pistil The ovule-bearing organ of a seed plant.

Pod Any dry, dehiscent fruit; specifically a legume.

Polygamo-dioecious Polygamous but chiefly dioecious.

Polygamous Bearing unisexual and bisexual flowers on the same plant.

Prairie A tract of grassland with no trees.

Prickle (prickly) Sharp outgrowth of the bark or epidermis.

Prostrate Lying flat upon the ground.

Pubescence (pubescent) A covering of short, soft hairs.

Raceme A simple, enlongated stem with each flower pediceled.

Rachis The axis of a spike or raceme, or of a compound leaf.

Radial Developing around a central axis.

Radiate Spreading from or arranged around a common center.

Rare Not close together; scattered.

Ray Usually referring to the more or less strap-shaped flowers in the head of Asteraceae.

Reclining Bending or turning toward the ground.

Recurved Bent backward.

Reflexed Abruptly bent or turned downward.

Remotely Distantly spaced.

Reniform Kidney-shaped.

Repand With an undulating margin.

Resaca A former course or channel of a stream, commonly water-filled to form narrow oxbow or meandering lakes.

Resinous Producing resin.

Reticulate Forming a network.

Rhizomes An underground stem or rootstock, rooting and bearing buds at the nodes.

Rhombic Somewhat diamond-shaped.

Rigid Stiff.

Rosette A crowded cluster of radiating leaves appearing to rise from the ground.

Saline Containing a salt.

Samara An indehiscent, winged fruit.

Scale (scales) A leaf much-reduced in size or an outgrowth of the epidermis.

Scandent Climbing.

Scant Not enough in size or quantity.

Scrub Stunted or densely packed bushes.

Sessile Without a stalk.

Shoot New stem of a shrub or tree.

Shrub (shrublet) A woody plant of smaller proportions than a tree, which usually produces several branches from the base.
Simple Unbranched, as a stem or hair; uncompounded, as a leaf.
Sinuate With a strongly wavy margin.
Spatulate Gradually narrowed downward from a rounded top.
Species A taxonomic category including closely related, morphologically similar individuals that interbreed and produce fertile offspring.
Spherical Round.
Spike An elongated rachis of sessile flowers or spikelets.
Spikelet A secondary spike.
Spine A sharp-pointed, stiff, woody body, arising from below the epidermis.
Spinescent More or less spiny, spine-tipped.
Spirally Winding, coiling, or circling around a center and gradually receding from or approaching it.
Sprawling To spread irregularly.
Spur A short, compact twig with little or no internodal development.
Stamen The male organ of the flower that bears the pollen.
Stem The organ of vascular plants that usually developes branches and bears leaves and flowers.
Stigma The part of the pistil of a flower that receives the pollen grains and on which they germinate.
Stomate (stomata) A small breathing pore or aperture in the epidermis of a plant.
Straggling To grow with long irregular branches.
Style Upward extension of the ovary terminating with the stigma.
Sub- Latin prefix meaning under or below.
Subshrub A perennial plant with the lower portions of stems woody and persistent.
Substrate The substance or base on which an organism grows.
Succulent Fleshy, thick, and juicy.
Sucker A stem originating from the roots or lower stem.
Suture A groove denoting a natural union.
Symmetrical Regularly, balanced arrangement on opposite sides of a line or plane, or around a center or axis.

Taproot The main or primary root.
Terete Round in cross section.
Ternate Consisting of or arranged in threes; trifoliolate.
Throat The orifice of a gamopetalous corolla; the expanded portion between the limb and the tube proper.
Tooth or toothed Having teeth; any small marginal lobe.
Transpiration Passage of water vapor outward, mostly through the stomata.
Transversed Lying or being across; place crosswise.
Tree A woody plant with one main stem.
Trifoliate Having three leaves.
Trifoliolate With three leaflets.
Trunk The main stem of a tree.
Twig A small shoot or branch.

Undulate With a wavy surface or margin.
Unisexual Flower having only stamens or pistils; of one sex.
Utricle A small, bladdery, one-seeded fruit.

Vaginate Loosely surrounded by a sheath.
Vegetative Having nutritive or growth functions as opposed to reproductive; nonwoody portion of plant.
Velvety Something like or suggesting velvet, as in softness, luster, etc.
Ventral Belonging to the lower face of an organ.
Vine A plant whose stem requires support.

Wart A glandular excrescence or hardened protuberance on plants.
Wavy Undulating or rolling, as a wavy terrain.
Whorl A ring of similar organs radiating from a node.

Xeric Characterized by or pertaining to conditions of scanty moisture supply.

L E A F   F O R M S

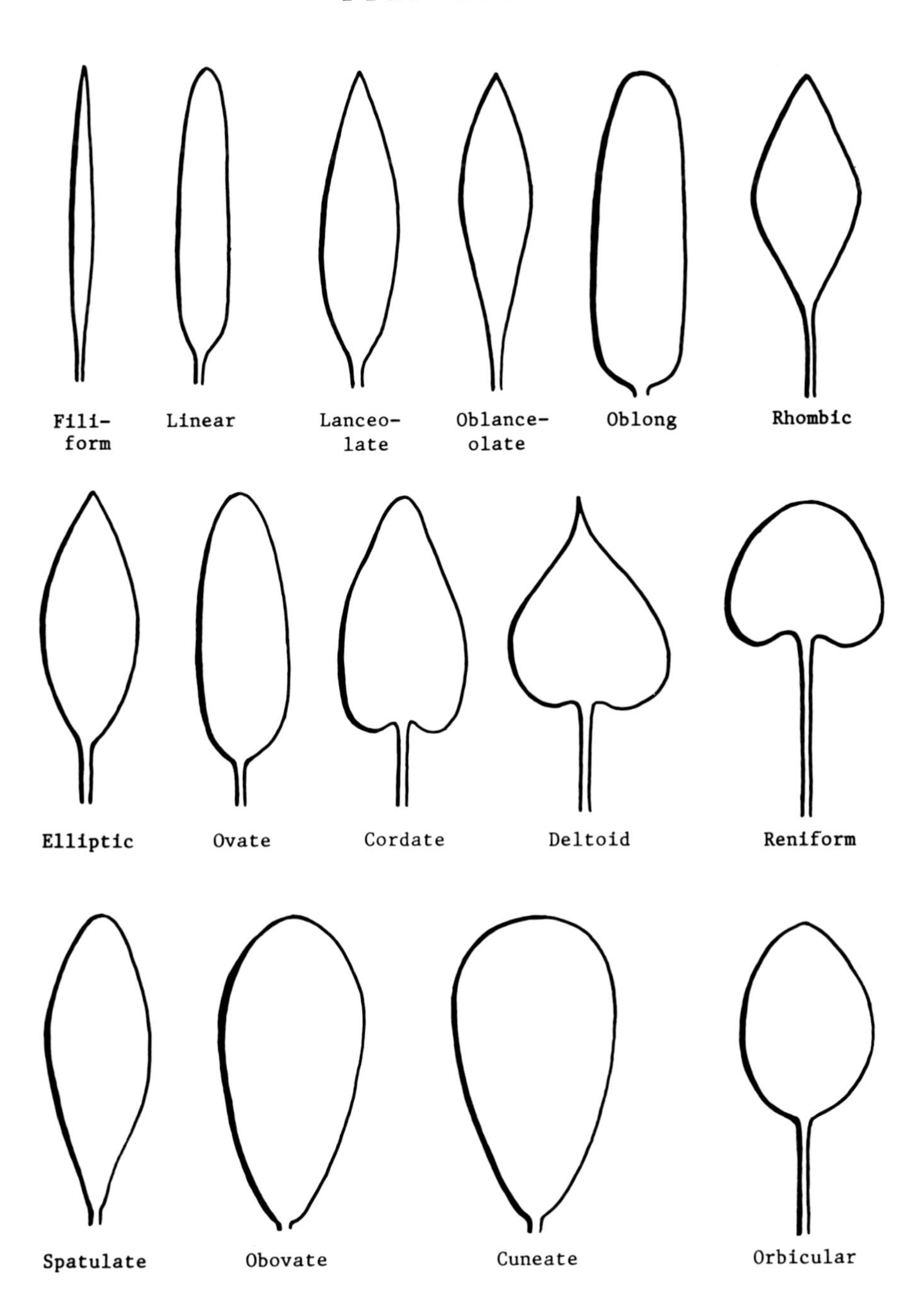
Fili-
form
Linear
Lanceo-
late
Oblance-
olate
Oblong
Rhombic
Elliptic
Ovate
Cordate
Deltoid
Reniform
Spatulate
Obovate
Cuneate
Orbicular

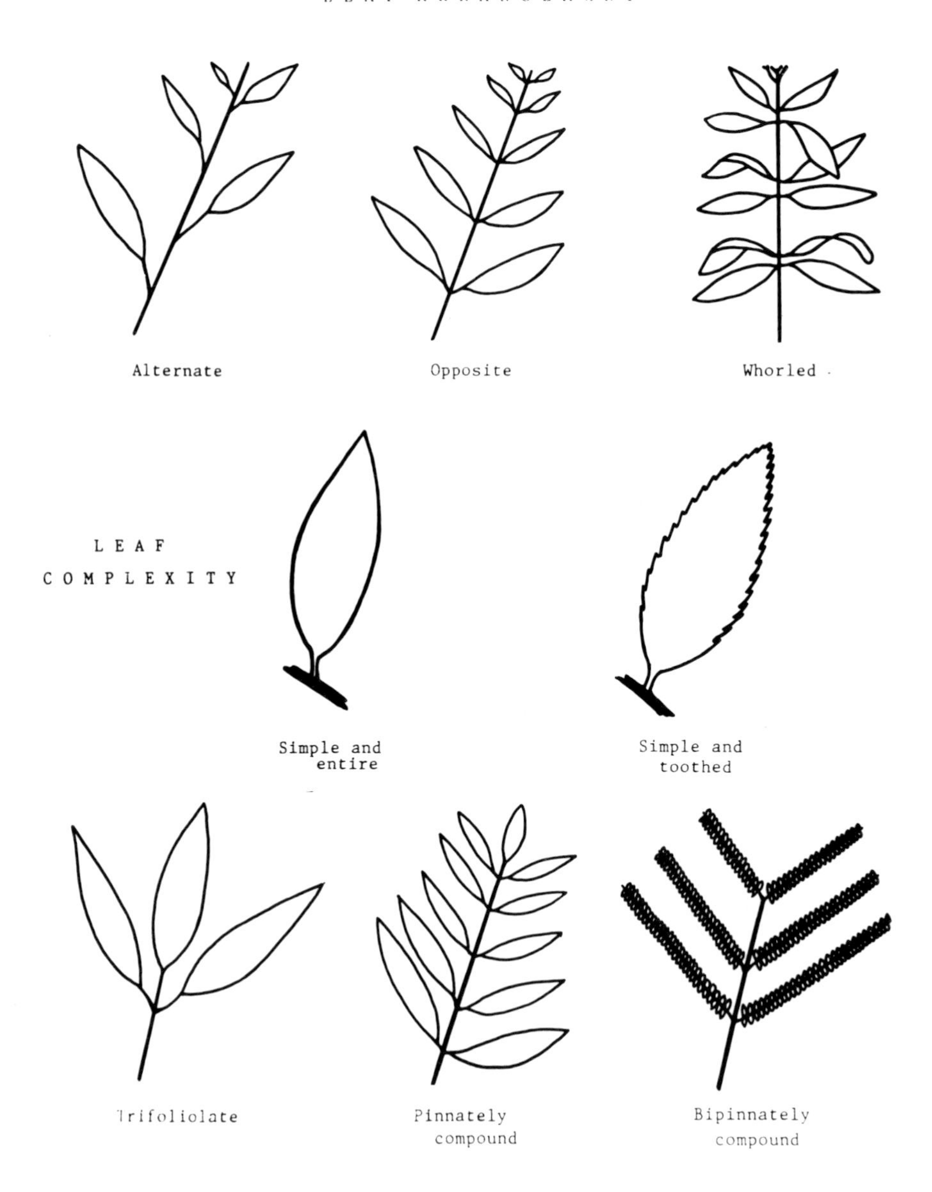
LEAF ARRANGEMENT
Alternate
Opposite
Whorled
LEAF
COMPLEXITY
Simple and
entire
Simple and
toothed
Trifoliolate
Pinnately
compound
Bipinnately
compound

LIST OF VERTEBRATES

Scientific and common names of birds are according to Rappole and Blacklock (1985), of mammals Jones *et al.* (1992).

BIRDS
American robin (*Turdis grayi*)
Cactus wren (*Campylorhynchus brunneicapillus*)
Cedar waxwing (*Bombycilla cedrorum*)
Great-tailed grackle (*Quiscalus mexicanus*)
Harris' hawk (*Parabuteo unicinctus*)
Mourning dove (*Zenaida macroura*)
Northern bobwhite quail (*Colinus virginianus*)
Northern cardinal (*Cardinalis cardinalis*)
Northern mockingbird (*Mimus polyglottos*)
Plain chachalaca (*Ortalis vetula*)
Pyrrhuloxia (*Cardinalis sinuatis*)
Rio Grande turkey (*Meleagris galopavo intermedia*)
Scaled quail (*Callipepla squamata*)
White-winged dove (*Zenaida asiatica*)
Whooping crane (*Grus americana*)

MAMMALS
Black-tailed jackrabbit (*Lepus californicus*)
Collared peccary or Javelina (*Tayassu tajacu*)
Common gray fox (*Urocyon cinereoargenteus*)
Coyote (*Canis latrans*)
Eastern fox squirrel (*Sciurus niger*)
Raccoon (*Procyon lotor*)
Southern plains woodrat (*Neotoma micropus*)
White-tailed deer (*Odocoileus virginianus*)

MEASUREMENT CONVERSION TABLE

EQUIVALENTS FOR CONVERSIONS FROM TEXT

Metric Unit		*U. S. System Equivalent*
Meter (m)	=	39.37 inches or 3.28 feet
Decimeter (dm)	=	3.937 inches
Centimeter (cm)	=	0.3937 inch
Millimeter (mm)	=	0.03937 inch

INDEX